Final Report

What Will Adaptation Cost? An Economic Framework for Coastal Community Infrastructure

June 2013

Eastern Research Group, Inc.

Written under contract for the
National Oceanic and Atmospheric Administration (NOAA)
Coastal Services Center

NOAA Coastal Services Center
(843) 740-1200
www.csc.noaa.gov

NOAA Coastal Services Center
LINKING PEOPLE, INFORMATION, AND TECHNOLOGY

Table of Contents

List of Tables

The Importance of Making Economically Informed Decisions

Global sea level has been rising over the past several decades and is expected to continue to rise in the decades to come. This shift poses a grave threat to coastal communities. Low-lying areas will experience more frequent inundation and shorelines will be more susceptible to increased erosion. Increased storm surges are already causing devastating floods. Sooner than most community leaders realize, some areas could be submerged so often or eroded so severely that people will be forced to abandon their homes and businesses. According to the National Ocean Economics Program, coastal communities account for nearly 60 percent of the nation's gross national product.[1] With so much of our nation's economy increasingly vulnerable to coastal hazards, communities need to consider future sea level rise in their decision-making.

Deciding on the best course of action, however, is not always easy. Issues are complex and there is a lot of uncertainty. Community leaders find themselves grappling with questions like:

- How will sea level rise and increased storm surge affect my community?

- What is the cost of doing nothing?

- What can we do to adapt?

- How can I determine the best adaptation strategy?

- How much will it cost to keep my community safe?

The purpose of this framework is to help communities begin to find answers to these difficult questions. By understanding the costs and benefits of different adaptation strategies, decision-makers can make more fully informed decisions that are fiscally responsible in the short and long terms. More importantly, economically informed decision-making will lead to safer, more responsible, economically sound communities. In the long run, the entire community benefits by investing in adaptation efforts: after a flood event, utilities will be restored quicker, stores and banks will be open sooner, children will return to school faster, and residents will be back at work with minimal disruption. Up-front investments can help ensure a successful future. By accounting for the full costs of inundation risks, leaders can make strategic choices about where, when, and how to make investments in adaptation responses to maximize benefits and minimize risk.

> "Going forward, I think we do have to anticipate these extreme types of weather. And we have to start to think about how we redesign [infrastructure] so this doesn't happen again...I don't think anyone can sit back anymore and say 'Well, I'm shocked at that weather pattern.' There is no weather pattern that can shock me at this point. And I think that has to be our attitude, how do we redesign our system and our infrastructure assuming that?"
> **—New York Governor Andrew Cuomo responding to Hurricane Sandy**

> "As we get more high tides and tides seem to get higher, and we get more of these storms and they seem to come with a little more fury, we get more and more water in our city as the days go by. So we are taking it very seriously ...and we are planning for it."
> **—Norfolk, Virginia, Mayor Paul Fraim**

---

[1] *State of the U.S. Ocean and Coastal Economies—2009.*

How to Use This Framework

This framework guides communities on how to evaluate options for adapting infrastructure to make it more resilient, reducing the effects of sea level rise (SLR) and high-water-level events such as storm surge or astronomical high tides. As a step-by-step process, it leads communities through a scenario-based approach to understand the full range of costs and benefits. Figure 1 illustrates the framework that communities can use to develop and investigate their own unique scenarios.

Figure 1: Framework for Making Informed Decisions

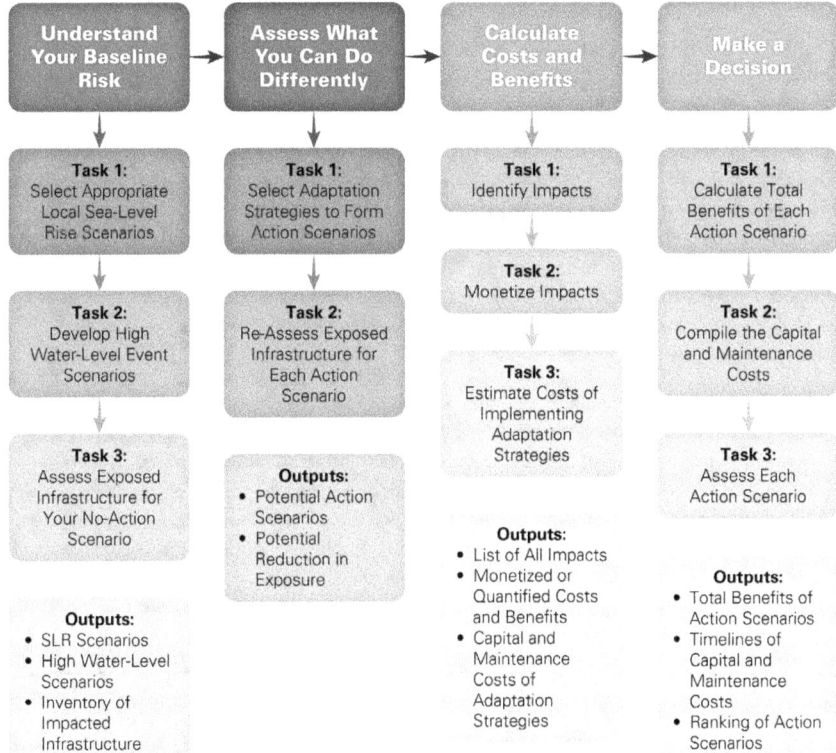

This framework is highly adaptable and can help any community make decisions about infrastructure investment. Communities can use it to assess the impact of inundation on their entire infrastructure, which is referred to as the holistic approach. Alternatively, if a community wants to focus on select infrastructure, such as a hospital or wastewater treatment plant, the framework allows for this priority infrastructure approach.

A significant impetus for developing this framework was the growing demand from communities for guidance to help protect publicly-owned infrastructure, such as roads, schools, and sewer systems. However, the holistic approach includes assessing costs and benefits to homes and businesses as well as public assets, and the adaptation strategies discussed in Chapter 2 and Appendix A can be used to protect both private property and public infrastructure. In fact, some of the adaptation strategies are traditionally employed to protect private, but can be viable options to protect public property as well. You can choose to use this framework to perform a more focused analysis of public infrastructure, or to take a broader look at your entire community.

> *"As storms such as Hurricane Sandy have shown, it is vital that we commit our resources and expertise to create a ready and resilient Maryland, by taking the necessary steps to adapt to the rising sea and unpredictable weather. In studying and planning for storms and climate change, we can ensure that our land, infrastructure, and most importantly our citizens are safe and prepared."*
>
> **—Maryland Governor Martin O'Malley**

The framework can also help communities perform either an impact assessment or a risk assessment. An impact assessment involves assessing the impacts of just a few different water-level increase heights to develop a sense of the amount of damage to expect in certain scenarios compared to the financial cost of adaptation strategies. A risk assessment is a more resource-intensive analysis that multiplies the probability of each water-level increase by the value of the impact to generate an apples-to-apples comparison of the expected costs and the expected benefits of implementing adaptation strategies against coastal flooding. The details of each type of assessment are discussed in task 2 of the first chapter in this guide.

Finally, we encourage communities to assemble a team to effectively use this framework, as several of its steps require specialized expertise or training. Identifying qualified experts and engaging stakeholders is an important part of the process. Your team should include economists, planners, GIS experts, land use attorneys, civil engineers, and inundation modeling experts if available; it can also include utility supervisors and emergency managers. If you do not have the needed expertise in-house for all the tasks, seek support from outside resources such as local universities, nonprofit organizations, or consultants. We recommend that communities engage other stakeholders and residents to participate in the decision process. This guide can also be used to help you craft an effective request for proposals to get necessary consultant support.

Protecting Against Sea Level Rise: New York City

When hazard planning includes informed decisions, successes can follow. In 2007, New York City began working to prepare for the impacts of climate change as part of PlaNYC—a comprehensive planning effort to address every aspect of the city's physical infrastructure. The effort recognized that adapting to SLR must be a key part of city policy. Accordingly, New York took action to address the most pressing risks; areas of the city were rebuilt, rezoned, and refashioned to better weather those threats.

When Hurricane Sandy hit the city in October 2012, many of these changes helped prevent property damage. Governors Island experienced peak flooding of almost 13 feet above sea level and saw 5 feet of water come over the southern seawall, but a 30-acre park there came through the storm nearly untouched because its design accounted for hazards like Sandy. The new Brooklyn Bridge Park was also designed with the assumption that it would flood—and was open to the public just five days after the storm.

Chapter 1: Understand Your Baseline Risk

This chapter helps you understand your community's baseline risk for coastal flooding resulting from high-water-level events and SLR. You will explore the exposure of your infrastructure to a range of water-level increases resulting from high-water-level events such as storm surge and astronomical high tides combined with SLR. The flood maps that you create will serve as a valuable tool to illustrate what might happen to your community if no action is taken.

Case Study: Boston, Massachusetts

In Boston, researchers used publicly available information from the U.S. Army Corps of Engineers and the Intergovernmental Panel on Climate Change to predict high-water-level event scenarios in 2050. They estimated that a 10-year storm in 2050 would bring the same height of water as a 100-year storm now, because of the compounded effects of SLR. In other words, the impacts of a storm that has a 1 percent chance of occurring now would equal those of a storm with a 10 percent chance of occurring in 2050.

Task 1:
Select Appropriate Local Sea-Level Rise Scenarios

Task 2:
Develop High Water-Level Event Scenarios

Task 3:
Assess Exposed Infrastructure for Your No-Action Scenario

Outputs:
- SLR Scenarios
- High Water-Level Scenarios
- Inventory of Impacted Infrastructure

Expertise required to complete Chapter 1:
- GIS analysts to help your team use spatial data in Task 3

Key resources referenced in Chapter 1 include:
- NOAA's *Global Sea Level Rise Scenarios for the United States National Climate Assessment*
- USACE Circular "Sea-level Change Considerations for Civil Works Programs"
- NOAA's Sea Levels Online
- NOAA's Extreme Water Levels map
- NOAA's Sea Level Rise and Coastal Flooding Impacts Viewer
- NOAA's Inundation Analysis Tool
- NOAA's *Mapping Coastal Inundation Primer*
- NOAA's Coastal LIDAR
- NOAA's Vertical Data Transformation
- USGS' National Elevation Dataset
- FEMA's Hazus-MH Flooding Model
- The New England Environmental Finance Center's Coastal Adaptation to Sea level Rise Tool

Task 1: Select Appropriate Local Sea Level Rise Scenarios

Objective: Use existing global and local projection data to select appropriate local SLR scenarios.

Process to complete this task:

Step 1: Identify a set of SLR curves that you will use for your analysis. It is possible that area researchers have developed localized SLR curves based on downscaled climate models, so you should check with area universities or the NOAA Regional Integrated Sciences and Assessments program closest to your community. For most locations, however, two recently published resources provide global SLR curves that can be used to develop your local SLR scenarios. This framework outlines how to use these two resources to generate local SLR scenarios. If you choose to use a different set of curves, you will need to apply quadratic equations for those curves to determine rates of SLR for different time periods.

Key considerations for this task:

- Are you considering a wide enough range of SLR scenarios? A low scenario based on historic SLR trends may make sense for short-term planning, but intermediate or high scenarios based on recent science showing that SLR is accelerating may be more appropriate to consider for longer-term planning (e.g. when planning for infrastructure meant to last multiple decades.)

- Did you project your local SLR scenarios far enough into the future? At this point, you have not looked at any adaptation strategies and their expected period of effectiveness. Consider projecting out to 2100 for now; you can shorten this horizon in a later task.

One available resource is U.S. Army Corps of Engineers (USACE) Sea-level Change Considerations for Civil Works Programs,[2] which presents three SLR curves: the low curve represents the historic or baseline rate, and the intermediate and high curves are based on science from the National Research Council and the Intergovernmental Panel on Climate Change. Option A in Step 3 explains how to use these curves to generate local SLR projections.

Another appropriate resource is NOAA's Global Sea Level Rise Scenarios for the United States National Climate Assessment. It is a multi-agency report of scientific literature on global SLR. Released in December of 2012, it presents four global SLR scenarios and notes the physical changes that would need to take place for each SLR scenario to be a reality (see Table 1.1). Options B and C in Step 3 of this chapter outline a simplified and more advanced methodology for using these curves to generate local SLR projections.

---

[2] This circular expires in September of 2013 but will likely be replaced by an updated version

Table 1.1: Global SLR Scenarios from *Global Sea Level Rise Scenarios for the United States National Climate Assessment*

SLR Scenario	Physical conditions scenario is based on	SLR by 2100 (ft)*
Lowest	Based on historical rates of observed sea level change	0.7
Intermediate-low	Based on projected ocean warming	1.6
Intermediate-high	Based on projected ocean warming and recent ice sheet loss	3.9
Highest	Based on ocean warming and the maximum plausible contribution of ice sheet loss and glacial melting	6.6

* Using mean sea level in 1992 as a starting point.

Step 2: Choose a few years along the planning horizon to evaluate local SLR. One approach would be every 30 years (2040, 2070, 2100), but there is no ideal planning horizon; timeframes of interest will be specific to your community, although you may want to consider performing this step out to 2100 to observe the effects of SLR over time. This will not limit you to a particular planning horizon in performing the cost-benefit analysis.

Step 3: Calculate local SLR increases for the years you chose, compared to the first year of your analysis. This involves incorporating local projection information into the global SLR curves, and using quadratic equations to determine the amount of rise for different years along the curves. Incorporating local information is necessary because different areas of the country have varying amounts of subsidence or uplift, and there are also differences due to ocean basin trends.

Option A: Use the three USACE curves to generate local SLR scenarios by year

Use the USACE on-line Sea Level Change Calculator to generate projections in five year increments. If you know the rate of subsidence or uplift for your area of coastline, you can enter this value into the calculator in millimeters. If you do not know this information, select the NOAA tide gauge closest to your location to generate an approximate subsidence rate. The calculator assumes a start date of 2010 for calculating values, but you may adjust this if desired. Use the results from the online calculator to fill out a table showing SLR rates for the time periods you selected in Step 2. Table 1.2 shows an example of what this table might look like, using St. Petersburg, Florida as an example. The values were generated using the online calculator by selecting the St. Petersburg tide gauge.

Table 1.2: St. Petersburg, Florida SLR Increase Scenarios by Year, Based on the USACE Curves.

SLR Scenario	Feet Above 1992* Sea Level		
	2040	**2070**	**2100**
High	0.93	1.97	5.25
Intermediate	0.67	1.21	2.93
Low	0.41	0.62	1.27

*The on-line Sea Level Change Calculator produces the amount of predicted sea level change from 1992 forward. 1992 corresponds to the midpoint of the current National Tidal Datum Epoch of 1983-2001.

Option B: Use the four curves in *Global Sea Level Rise Scenarios for the United States National Climate Assessment* to generate local SLR scenarios by year – Simplified methodology

A relatively simple way to adjust the global SLR curves to produce local scenarios is to factor in the difference between the global mean sea level trend and the trend in your location, which may be greater or lesser than the global rate depending on local vertical land movement (subsidence or uplift) and regional ocean basin trends. Use the following steps to generate values for the four scenarios (lowest, intermediate-low, intermediate-high, and high.) Again, St. Petersburg, Florida is used as an example, and Table 1.4 illustrates the final output.

1. Determine the lowest rate (in/year) of SLR increase using the NOAA Sea Levels Online mean sea level trend from the water level gauge closest to your community. Assume this to be a constant increase, and use 1992 as the starting point for your calculations; use it to calculate your "lowest" projection for the years you chose in the previous step of this process. For St. Petersburg, Florida, the mean sea level trend is 2.36 mm/year, which converts to 0.093 in/year. The rise for 2040 will be 4.46 in (0.093 in/year x [2040-1992]), for 2070 will be 7.25 in (0.93 in/year x [2070-1992]), and for 2100 will be 10.04 in [0.093 in/year x [2100-1992]). Insert these values in the row for the "lowest" scenario in Table 1.4. (Note that Table 1.4 is in feet so these values have been converted from inches to feet.)

2. Use the quadratic equations in Section 4.3 of the *Global Sea Level Rise Scenarios for the United States National Climate Assessment* report (shown in Equation 1.1 below) to calculate global SLR associated with the "intermediate-low," "intermediate-high," and "highest" SLR scenario for the years you have chosen to analyze. An example of calculating these equations is provided below, and Table 1.3 provides an example of the values (in feet) for the three time periods of 2040, 2070, and 2100.

Future global mean SLR is represented by the following quadratic equation:

$$E(t) = 0.0017t + bt^2 \qquad \text{(Equation 1.1)}$$

Where:
t = years, starting in 1992
b = constant value of 1.56E-04 for the Highest Scenario, 8.71E-05 for the Intermediate-High Scenario, and 2.71E-05 for the Intermediate-Low Scenario[3].
E(t) = eustatic SLR, in meters, as a function of t.

Using the highest scenario in the year 2070 as an example, t equals 78 (i.e. 2070 – 1992), and the equation would be:

$$E(t) = (0.0017 \times 78) + [(1.56 \times 10^{-4}) \times (78^2)]$$

This yields a result of 1.08 meters, which converts to 3.54 feet.

Table 1.3: Global SLR Increase Scenarios by Year Based on the Global Sea Level Rise Scenarios for the United States National Climate Assessment Curves

Global SLR Scenario	Feet Above 1992 Sea Level		
	2040	2070	2100
Highest	1.4	3.5	6.6
Intermediate-high	0.9	2.2	3.9
Intermediate-low	0.5	1.0	1.6

3. Adjust the global "intermediate-low," "intermediate-high," and "highest" values calculated in the previous step to account for the difference between your local mean sea level trend and the global mean SLR average (0.067 in/year.) For example, St. Petersburg's local SLR rate of 0.093 in/year is 0.026 in/year above the global average. So for the year 2040, 1.25 inches need to be added to each of the three scenarios (0.026 in/year x [2040-1992]). For the year 2070, 2.03 inches need to be added (0.026 in/year x [2070-1992]), and for the year 2100, 2.81 inches need to be added (0.026 in/year x [2100-1992]). Insert the adjusted numbers into the "highest," "intermediate-high," and "intermediate-low" rows of table 1.4.

Table 1.4: Example Local SLR Increase Scenarios by Year Based on the St. Petersburg, Florida tide gauge and the *Global Sea Level Rise Scenarios for the United States National Climate Assessment* Curves

Local SLR Scenario (St. Petersburg, FL)	Feet Above 1992 Sea Level		
	2040	2070	2100
Highest	1.5	3.7	6.8
Intermediate-high	1.0	2.4	4.1
Intermediate-low	0.6	1.2	1.8
Lowest	0.37	0.60	0.84

---

[3] *Global Sea Level Rise Scenarios for the United States National Climate Assessment (2012)*

Option C: Use the four curves in *Global Sea Level Rise Scenarios for the United States National Climate Assessment* to generate local SLR scenarios by year – Advanced methodology

For the greatest accuracy, follow the detailed guidance in sections four and five of *Global Sea Level Rise Scenarios for the United States National Climate Assessment* to calculate local SLR estimates.

1. Use Table 1.5 below as a template to help you generate your estimates for the year 2100. As the table footnotes indicate, *Global Sea Level Rise Scenarios for the United States National Climate Assessment* provides more information about developing values for Table 1.5.

Table 1.5: Template for Calculating Local SLR for 2100

Contributing Variables	SLR Scenarios			
	Lowest Scenario	Intermediate Low Scenario	Intermediate High Scenario	Highest Scenario
Global mean SLR from 1992 to 2100 (ft)*	0.7	1.6	3.9	6.6
Vertical land movement (subsidence or uplift) (ft)**				
Ocean basin trend (from tide gauges and satellites) (ft)***				
Total Relative Sea Level Change (ft)				

* Equations from Section 4.3 of *Global Sea Level Rise Scenarios for the United States National Climate Assessment* can be used to calculate scenarios of sea level change over desired period and to populate the global mean SLR term in the first row.

** This row can be populated using, in part, the information found in Sections 5.1 of *Global Sea Level Rise Scenarios for the United States National Climate Assessment*.

*** This row can be populated using, in part, the information found in Sections 3.1, 3.2, 3.3 and 5.3 of *Global Sea Level Rise Scenarios for the United States National Climate Assessment*

2. Once you have generated values for 2100, create additional tables for your selected years between now and 2100. Use the quadratic equations in Section 4.3 of the *Global Sea Level Rise Scenarios for the United States National Climate Assessment* to complete the top row of each table. (There are equations for the "intermediate-low," "intermediate-high," and "highest" scenarios; for the "lowest" scenario, multiply the global mean SLR average (0.067 in/year) times the number of years since 1992.)

3. Once you have calculated local SLR for each of your selected time periods, create one final table with the total relative sea level change for each time period and each of the four SLR scenarios. This final table would follow the format of Table 1.4 shown above.

Task 2: Develop High-Water-Level Event Scenarios

Objective: Review historical data to select a range of high-water-level events. Integrate these data with SLR scenarios to select several water-level height increases that you will use for the basis of your risk or impact assessment.

Process to complete this task:

Step 1: Using the NOAA Extreme Water Levels map, find the "Exceedence Probability Curves" for the available city closest to your community. Use the bolded curve on the graph to determine the water-level increase relative to the mean higher high water (MHHW) level for any storm intensity from the 1-year storm to the 200-year storm. Consider selecting at least the 1-year, 2-year, 10-year, 30-year, 100-year, and 200-year storm in developing this initial table. (Note that "10-year storm" refers to a storm that is projected to occur once every 10 years. That is, such a storm has a 10 percent chance of occurring in any given year—it could happen more than once.) See the first two columns of Table 1.6, which show the storm-type and present water-level increases, for a sample of how this table might look. This table presents two SLR scenarios; however, your community may choose to examine more or fewer scenarios in your analysis.

Key considerations for this task:

- Did you consider extreme water-level increases? Considering that future climate change could exacerbate the strength and frequency of storms and global mean sea level is projected to increase at an accelerated pace, it is very important that a community develop informed adaptation strategies while considering the possibility of catastrophic water-level increases.

- Do you have the resources to perform a risk assessment? A risk assessment provides an apples-to-apples comparison of your costs and benefits; however, it requires assessing the impacts of significantly more water-level increases than an impact assessment.

What is a 100-year storm?

A 100-year storm has a 1-in-100—or 1 percent—chance of occurring in a given year. This does not guarantee that it will occur exactly once every 100 years, nor does it preclude it from happening twice in a year. Here are some example storm types and their associated annual probability of occurring:

- 200-year storm: 0.005 percent
- 100-year storm: 1 percent
- 30-year storm: 3.3 percent
- 10-year storm: 10 percent
- 2-year storm: 50 percent

Table 1.6: Sample Table of Water-Level Increases by Year (feet)

Scenario/Year	Present*	2040	2070	2100
Intermediate-Low SLR Scenario (3-Foot Rise by 2100)				
200-year storm	12	13	14	15
100-year storm	10	11	12	13
30-year storm	8	9	10	11
10-year storm	5	6	7	8
2-year storm	4	5	6	7
1-year storm	3	4	5	6
Highest SLR Scenario (6-Foot Rise by 2100)				
200-year storm	12	14	16	18
100-year storm	10	12	14	16
30-year storm	8	10	12	14
10-year storm	5	7	9	11
2-year storm	4	6	8	10
1-year storm	3	5	7	9

* These data are for illustrative purposes only and do not represent any particular exceedance probability curve.

Step 2: Expand your table to include your SLR scenarios from Chapter 1, Task 1. Add all the scenarios you developed for your chosen years. Table 1.6 shows what this might look like. The integrated water-level increase is the sum (it should be adequate to round to the nearest foot) of the water-level increase associated with a certain storm type and the water-level increase in your SLR scenario for that given year.

Step 3: Decide whether to perform an impact assessment or a risk assessment. The steps in either case are similar, but vary according to your decision—be sure to pay attention when specific steps for your assessment type are called out in each task. Risk assessment is significantly more resource intensive, but gives you a much more informative comparison of your costs and benefits.

When performing an impact assessment, you will assess three or four water-level increases as they relate to infrastructure. However, you could assess over 10 water-level

> **What is the difference between a risk assessment and an impact assessment?**
>
> An **impact assessment** calculates the dollar value of flooding damage for your selected water-level increases. A **risk assessment** takes the analysis one step further by multiplying the probability of each water-level increase occurring by the value of the impact to yield a direct comparison of expected costs and expected benefits.

increases to adequately perform a risk assessment. An impact assessment could give you enough information to assess what combination of events could justify the costs of implementing your adaptation strategies. A risk assessment factors in the probability of the storm and will allow you to directly compare your expected total costs and benefits, giving you a clearer answer as to whether your adaptation strategies are worthwhile. The risk assessment does not factor in the probability of an SLR scenario; there are no probabilities associated with each SLR scenario. That is, your community will select an SLR scenario as a given assumption, and perform a risk assessment for each SLR scenario you choose to assess. If you do choose an impact assessment and (on reaching the last chapter of this framework) find that you need more detailed cost and benefit data to make your decision, you can always go back and replicate many of these tasks for more water-level increases to perform a risk assessment.

Step 4: Select the water-level increases you will assess for either your impact or risk assessment. The specific process you will undertake for each assessment type is outlined below.

Option A: impact assessment

Select three or four water-level increases that represent a range of the values in your table. Consider choosing a low, medium, high, and perhaps catastrophic water-level increase. A low water-level increase may be, for example, an astronomical high tide, a 1-year storm, or SLR only. A catastrophic water-level increase may be, for example, the 200-year storm in 2100 under the "highest" SLR scenario or the water-level increase associated with the highest surge that has ever been experienced in your region of the country—Sandy, Katrina, etc. Create a table showing the selected water-level increases (see Table 1.7 for an example) and the associated storm type and SLR scenario combinations represented by those increases.

Table 1.7: Sample Table of Selected Water-Level Increases for the Impact Assessment

Water Height*	What Does This Height Represent?
3 feet	Present 1-year storm, intermediate-low SLR-only in 2100, high SLR only around 2055.
8 feet	30-year storm in present, 10 year-storm in 2100 under intermediate-low SLR scenario or 2055 under highest SLR scenario, 2-year storm in 2070 under highest SLR scenario, 1-year storm in 2085 under high SLR scenario.
12 feet	200-year storm in present, 100-year storm in 2070 under intermediate-low SLR scenario or 2040 under highest SLR scenario, 30-year storm in 2070 under highest SLR scenario.
18 feet	200-year storm in 2100 under highest SLR scenario.

* These data are for illustrative purposes only and do not represent any particular exceedance probability curve.

Option B: risk assessment

Choose one or more SLR scenarios for which you will assess every water-level increase in the table. For example, for the sample water-level increases in Table 1.6, you will assess water-level increases of 3, 4, 5, 6, 7, 8, 9, 10, 11, 12, 13, 14, and 15 feet if you choose to assess only the intermediate-low scenario. If you choose to assess the highest scenario as well, you will also need to select water-level increases of 16 feet and 18 feet. In many cases, including this sample case, there is only a small increase in the number of water-level increases that you will need to assess as you consider more SLR scenarios.

Task 3: Assess Exposed Infrastructure for Your No-Action Scenario

Objective: Identify at-risk infrastructure and land area for each water-level increase.

Process to complete this task:

Step 1: Use water-level increases developed in the previous task and the NOAA Sea Level Rise and Coastal Flooding Impacts Viewer (where available) to make a reconnaissance-level assessment of the exposed land and infrastructure in your local community. Note that this step is not necessary to the analysis; when available, though, it can provide a useful visual of flooding in your community for 1- to 6-foot water-level increases.

Step 2: Determine whether you are interested in assessing the impacts to specific elements of your infrastructure such as a wastewater treatment plant, utility, or hospital (the priority infrastructure approach), or assessing the impacts to your

Key considerations for this task:

- Are you interested in assessing the impacts to all infrastructure in your community (holistic approach), or will you be choosing one or two priority infrastructure elements to assess (priority infrastructure approach)?

- What is the complexity of your flooding map, and what are its limitations? Some are simplifications that just consider elevation and would show an area flooding even with a mountain shielding it from storm surge. Others account for this by evaluating how the connectivity of adjacent land parcels affect the flood waters (e.g., hills that prevent the tide from coming further inland. Does your modeling account for wave action?

- What is the resolution digital elevation map data? This will impact the uncertainty of your flood modeling.

Experts required for this task:

- GIS analysts to help your team use spatial data in Step 3.
- Inundation modeling expert, if available, to assist with all three steps.

community as a whole (the holistic approach). Generally, the tasks will be the same for either approach; however, certain tools and methodologies in this framework are more tailored toward a specific approach. Although the steps in each task will typically be the same either way, be sure to note cases in which a task calls out specific steps for your approach type.

Step 3: Determine the exposure, including the depth of flooding relative to the bottom of your infrastructure, for the infrastructure you will be considering in your analysis. While it is not critical to performing this step, it is helpful to understand how much of the water-level increase is due to permanent inundation from sea-level rise as that will impact your community differently from temporary flood from a high-water-level event such as storm surge or astronomical high tides. For both the holistic and priority-infrastructure approach, you might find that flooding levels are relative to one point of reference, typically MHHW, and the elevation data sets are relative to another point of reference, typically North American Vertical Datum. Use NOAA's Vertical Data Transformation tool or another method to convert your data to a common reference point if this is the case. This step varies for the holistic and priority infrastructure approach.

Option A: holistic approach

This step of the process involves creating inundation (flooding) maps and will require GIS or inundation mapping experience. See the NOAA *Mapping Coastal Inundation Primer* for more information about different approaches to developing these models, data needs for each approach, and data resources. State and local GIS staff members can be a great resource to help you get elevation data and mapping tools; also useful are NOAA's Coastal LIDAR, NOAA's Digital Elevation Model Discovery Portal, USGS' National Elevation Dataset, USGS' Center for Lidar Information Coordination and Knowledge (CLICK), and the Open Topography portal—among others. Alternatively, consider using comprehensive tools such as the FEMA Hazus MH flooding tool and COAST. These tools also require expertise, but they are also programmed to monetize (or assign a dollar value to) the damage costs associated with water-level increase inputs. You can read more about them in **Appendix B.**

Option B: priority infrastructure approach

You can typically complete this step by determining the elevation of the bottom of your infrastructure relative to MHHW using elevation data such as a digital elevation map, and determining the depth of flooding associated with each water-level increase relative to MHHW selected in the previous task. For example, if your coastal infrastructure is 6 feet above MHHW, and your selected water-level increase is 14 feet above MHHW, you would assume approximately 8 feet of flooding relative to the base of your infrastructure.

Chapter 2: Assess What You Can Do Differently

Once your community understands the severity of your baseline risks from coastal flooding, the next step is to explore how to adapt to these risks. In this chapter, you will develop action scenarios, which will be composed of one or more adaptation strategies to make your community more resilient to coastal flooding. You can assess each action scenario to determine how it alters the severity of flooding in your community compared to your baseline case. This chapter will set the stage for you to apply dollar values to the damage of your no-action and action scenarios in order to arm your community with the information it needs to make fiscally and socially responsible decisions about implementing adaptation strategies.

> **Task 1:**
> Select Adaptation Strategies to Form Action Scenarios

> **Task 2:**
> Re-Assess Exposed Infrastructure for Each Action Scenario

Case Study: Hampton Roads, Virginia

In Hampton Roads, the nation's 35th largest metropolitan area, the cities of Norfolk and Virginia Beach face starkly different decisions in selecting resilient infrastructure and adaptation strategies in response to higher anticipated sea level rises than any other Atlantic Coast metropolitan area. Norfolk is the financial, medical, and cultural center of the region, with a downtown waterfront of about 7 miles. Hard infrastructure projects, such as seawalls and levees, protect its major downtown assets and will likely be an important component of future adaptation. Virginia Beach, however, relies primarily on its tourism industry made possible by its miles of white sand beaches. Investments in beach nourishment have countered erosion while maintaining the natural aesthetic qualities of the beach.

> **Outputs:**
> - Potential Action Scenarios
> - Potential Reduction in Exposure

Expertise required to complete Chapter 2:
- GIS analysts to help your team use spatial data in Task 2
- Land use attorney and planners to help select adaptation strategies in Task 1
- Civil engineers to help evaluate adaptation strategies in Task 2

Key resources referenced in Chapter 2 include:
- Adaptation strategy fact pages in **Appendix A** of this framework
- NOAA's *Adapting to Climate Change: A Planning Guide for State Coastal Managers*
- EPA's *Synthesis of Adaptation Options for Coastal Areas*
- UNEP's *Technologies for Climate Change Adaptation*
- UNESCO's Hazards Awareness and Mitigation in Integrated Coastal Area Management
- Climate Tech Wiki

Task 1: Select Adaptation Strategies

Objective: Develop action scenarios for flooding adaptation strategies. An adaptation strategy is an individual measure, such as elevating development, that reduces the impacts of flooding. An action scenario is a compilation of one or more of these strategies.

Process to complete this task:

Step 1: Review Table 2.1 to identify possible adaptation strategies that could help reduce the exposure of infrastructure to coastal flooding in your community. Use **Appendix A,** which includes a fact sheet about each of the adaptation strategies in Table 2.1, to learn more about the cost, effectiveness, implementation obstacles, and other considerations about each measure. Investigate other measures not listed in the table that other communities have considered

Key considerations for this task:

- Do your adaptation strategies directly respond to your risks?

- Do your action scenarios range in cost and effectiveness? A wide range of scenarios can establish the foundation for constructive discussions.

- What is the period of effectiveness of your adaptation strategies? Many solutions are only temporary and need to be combined with plans for managed retreat in the future.

- Can a combination of smaller adaptation strategies produce results at least as good as a more expensive adaptation strategy goal?

Experts required for this task:
- Land use attorney and planners to help select adaptation strategies in Steps 1 and 2.

or implemented, like moving expensive equipment to higher ground in a hospital or other priority infrastructure. While Table 2.1 and Appendix A provide some general, introductory information, there are other factors to consider as you select adaptation strategies that include:

- The costs of strategies can vary significantly depending on the unique characteristics of your community.

- Non-economic factors such as legal challenges or public outreach needs can increase the resources needed to implement a strategy.

- The lifespan and effectiveness of any project will depend on the severity of future events.

Keep in mind that some of these adaptation strategies may be more geared to an entire community and others are more applicable to protecting priority infrastructure.

Step 2: Based on your list of possible adaptation strategies, develop one or more action scenarios. Each action scenario should consist of at least one adaptation strategy; you may want to combine strategies to enhance the effectiveness of the action scenario.

Table 2.1: How Adaptation Strategies Reduce the Impacts of Coastal Flooding

Adaptation Strategy	How This Adaptation Strategy Changes the Impacts of Coastal Flooding
Managed Retreat Policies	
Transfer of development rights (TDR)	Encourages future development to be located out of harm's way.
Purchase of development rights (PDR)	Encourages future development to be located out of harm's way.
Rolling easements	May discourage future development from being located in harm's way. Can lead to removal of existing development from harm's way as shorelines move inland.
Fee-simple acquisition (buyout)	Prevents new development from being located in harm's way and/or removes development currently in harm's way.
Infrastructure relocation	Relocates the infrastructure out of harm's way.
Tidal Management	
Storm-surge barriers	Prevents higher water from traveling through inlets or into estuaries up to a certain water-level increase.
Engineered Barriers	
Levees and dikes	Prevents flooding up to a certain water-level increase.
Sea walls	Prevents flooding up to a certain water-level increase.
Beach nourishment	Prevents flooding up to a certain water-level increase.
Sandbagging	Prevents flooding up to a certain water-level increase.
Infrastructure Modification/Design	
Elevated development	Reduces the damage caused by flooding by raising the infrastructure above ground level.
Flood-proofing infrastructure	Reduces the damage caused by flooding.
Floating development	Prevents flooding to structure as the development rises with the water.
Floodable development	Prevents structural damage up to a certain height. May contain some water which can prevent flooding of other assets.
Movable buildings	Allows for relocating the infrastructure out of harm's way.
Drainage systems	Manages flood water to reduce damage.
Land Use Policy	
Preservation of open space	Prevents future development from being located in harm's way. Preserved open space may also absorb flood water and/or serve as a buffer during inundation events.
Zoning in vulnerable areas	Minimizes or prevents future development from being in harm's way or requires future development to be more resilient to flooding.

Table 2.1: How Adaptation Strategies Reduce the Impacts of Coastal Flooding

Adaptation Strategy	How This Adaptation Strategy Changes the Impacts of Coastal Flooding
Development fees in vulnerable areas	Can be used to pay for flood mitigation measures and may encourage future development to be located out of harm's way.
Green Infrastructure	
Wetlands	Absorb water to reduce the overall water-level increase, and dissipate wave and storm surge energy.
Mangroves	Reduce the wave power, typically resulting in a smaller storm surge and a slightly lower water-level increase.
Oyster and coral reefs	Reduce the wave power, typically resulting in a smaller storm surge and a slightly lower water-level increase.
Living dunes	Prevent flooding up to a certain water-level increase.
Barrier island restoration	Reduces the wave power, typically resulting in a smaller storm surge and a slightly lower water-level increase.

Task 2: Re-Assess Your Exposed Infrastructure for Your Action Scenarios

Objective: Identify your infrastructure and land area at risk for each action scenario.

Process to complete this task:

Step 1: Identify how each action scenario changes the coastal flooding impacts for each water-level increase identified in Chapter 1. Determine the impact of each action scenario. For example, does it move your infrastructure out of harm's way to completely prevent flood damage, reduce the frequency of inundation, prevent flooding up to certain water-levels, make the building more resilient against flooding, or combine some of these possibilities? See Table 2.1 and **Appendix A** for more information about how each flooding adaptation strategy reduces the impact of coastal flooding.

Key considerations for this task:

- Does your action scenario move infrastructure out of harm's way, reduce the level of flooding for a given water-level increase, or completely prevent flooding up to a certain water-level increase? The key to re-assessing the change in exposure, and ultimately the change in impacts, is to understand and make assumptions about how your action scenarios change coastal flooding and the resulting impacts compared to the no-action scenario.

Experts required for this task:

- GIS analysts to help your team use spatial data in Step 2.
- Civil engineers to help assess the effectiveness of your action scenario against coastal flooding. In Step 1.

Step 2: Re-assess the exposure of your infrastructure for each action scenario. This step's complexity will depend on how your action scenario changes the level or impacts of flooding.

If your action scenario prevents flooding up to a certain water-level increase:

You will eliminate impacts of flooding for water-level increases up to a certain point; however, once that point is exceeded, the flooding impacts will be similar to those in a no-action scenario. For example, beach nourishment may prevent flooding completely up to a 5-foot water-level increase; for a 10-foot increase, though, the flooding might be the same as in the no-action scenario. This will not require GIS modeling beyond what you did in Task 3 of Chapter 1. Rather, you will just need to note the water-level increases for which impacts are prevented and those for which impacts will be similar to the no-action scenario.

If your action scenario reduces the severity of flooding for a given water-level increase:

You will need to determine by how much the level of flooding changes. For example, if a wetland restoration project or series of projects reduce a 10-foot water level by 1 foot, create a new GIS flood map for 1 foot less of flooding or note that you have 1 foot less of flooding for any priority infrastructure.

If your action scenario moves your infrastructure out of harm's way:

You will not need to create any new flood maps, since the level of flooding associated with a given water-level increase does not change. When you assign dollar values to the impacts in the next step, the difference in damage between your no-action and action scenarios will be the avoided damage to this relocated infrastructure.

If your action scenario increases the resilience of your infrastructure:

You will not need to create a new flood layer. For example, if you flood-proof or elevate your infrastructure, the amount of damage to your infrastructure will be reduced for a given level of flooding. Though this will not require a new flood map, it will reduce the monetized value of damage to that infrastructure for a given level of flooding, so you will have to track which infrastructure elements are involved and change the depth-damage curve in Chapter 3. By isolating the changes in damage to these particular buildings, you can estimate the avoided costs.

Chapter 3: Calculate Costs and Benefits

After determining the exposure of your infrastructure in both no-action and action scenarios, you can identify and then monetize—or assign a dollar value to—the impacts of coastal flooding on your infrastructure or your community. These include direct impacts from coastal flooding as well as impacts, both positive and negative, from implementing your adaptation strategies. In practice, it will be too resource intensive to monetize or quantify all impacts, so you will need to consider some impacts qualitatively to avoid underestimating the value of impacts.

Case Study: Lower Fox River Basin, Wisconsin

In the Lower Fox River Basin, Resources for the Future estimated flood damage with Hazus-MH Flooding Model, a GIS-based model developed for the Federal Emergency Management Agency (FEMA) to help local officials and emergency planners estimate losses from floods, earthquakes, and hurricanes. They modeled baseline flood damage estimates for 10-year, 50-year, 100-year, and 500-year flood events in terms of total building, content, and inventory loss; business interruption loss; the number of moderately damaged buildings; truckloads of debris generated; the number of displaced households; and agricultural losses—all based on the current land-use patterns. The predicted total losses ranged from $47.5 million for a 10-year flood event to almost $109 million for a 500-year flood.

Task 1:
Identify Impacts

Task 2:
Monetize Impacts

Task 3:
Estimate Costs of Implementing Adaptation Strategies

Outputs:
- List of All Impacts
- Monetized or Quantified Costs and Benefits
- Capital and Maintenance Costs of Adaptation Strategies

Expertise required to complete Chapter 3:
- Experienced economists to help you identify and monetize the costs and benefits in Tasks 1 and 2
- GIS analysts to help your team use spatial data in comprehensive monetization tools in Task 2
- Engineers to help you estimate the costs of adaptation strategies in Task 3

Key resources referenced in Chapter 3 include:
- Monetization approaches in **Appendix B** of this framework
- Comprehensive tools including FEMA's Hazus-MH Flooding Model and the New England Environmental Finance Center's Coastal Adaptation to Sea Level Rise Tool (COAST)
- The U.S. Army Corps of Engineers' depth-damage functions
- Parametric software including RS Means, the RS Means Quick Cost Estimator, the U.S. Cost Success Estimator, and the Parametric Cost Engineering System

Task 1: Identify Impacts

Objective: Recognize and categorize the potential impacts of coastal flooding on your infrastructure and community and the impacts from implementing resilient infrastructure options.

Process to complete this task:

Key considerations for this task:
- Are there additional impacts not included in the "Comprehensive Listing of Impacts" tables below that are specific to your community?

- Do you have internal team members with basic economic understanding to generate a list of impacts or will you need an economist to review your list?

Experts required for this task:
- Experienced economists to help you identify impacts in Steps 1-3.

Step 1: Create a list of primary, secondary, and environmental impacts from flooding using Table 3.1 for input. Keep in mind that impacts will depend on the severity of flooding for certain action scenarios, and some impacts might not exist at all under certain action scenarios or water-level increases. However, for the purpose of this task, develop a single comprehensive list of all impacts that your community might face under any flooding scenario. If you are only assessing the impacts on priority infrastructure, create a similar list identifying the direct impacts to your infrastructure and resulting secondary impacts. You may want to provide a higher level of detail about the impacts such an inventory of specific content in the infrastructure damaged or components of the infrastructure that would be affected.

Step 2: List potential impacts of implementation, both positive and negative, for each adaptation strategy identified in your action scenario. See Table 3.2 for input. Make a separate list of impacts for each action scenario identified in Chapter 2. This step will be similar for both the priority infrastructure and holistic approach.

Step 3: Brainstorm whether your community needs to consider any other impacts, not listed in Tables 3.1 or 3.2. Add these to the two lists you developed. This step will be similar for both the priority infrastructure and holistic approach.

Table 3.1: List of Impacts from Flooding

Impact	Tool or Measurement Methodology						
	FEMA Hazus MH Flooding	COAST	Overlay Infrastructure-Level Economic Data on Flood-Depth Data	Use General Economic and Flood Data to Estimate Primary Damage	Non-Market Valuation Methodologies	Monetize Business Interruption Loss	Benefits Transfer
Primary Impacts							
Residential building damage*	•	•	•	•			•
Commercial building damage*	•	•	•	•			•
Damage to special facilities such as hazardous waste, wastewater treatment plants, landfills, and energy utilities*	•	•	•	•			•
Damage to essential facilities such as hospitals, fire stations, and schools*	•	•	•	•			•
Vehicle damage	•		•				
Building content loss*	•	•	•				•
Road damage					•		
Bridge damage	•						•
Railway damage							•
Crop loss	•						•
Loss of human life							•
Animal and livestock loss							•
Secondary Impacts							
Business interruption costs*	•					•	•
Debris cleanup	•						•
Emergency response	•						•
Evacuation costs	•						•
School hours loss					•		•
Rental income loss	•						•
Relocation costs	•				•		•
Displaced families	•				•		•
Roadway congestion					•		•

Table 3.1: List of Impacts from Flooding

Impact	Tool or Measurement Methodology						
	FEMA Hazus MH Flooding	COAST	Overlay Infrastructure-Level Economic Data on Flood-Depth Data	Use General Economic and Flood Data to Estimate Primary Damage	Non-Market Valuation Methodologies	Monetize Business Interruption Loss	Benefits Transfer
Anxiety and discomfort					•		•
Environmental Impacts							
Salinization of freshwater supply					•		•
Parks, campgrounds, and beaches destroyed					•		•
Wastewater intrusion					•		•
Erosion					•		•
Wetlands, mangroves, marshes, and estuaries destroyed					•		•
Other ecosystem service related costs					•		•

*   Most substantial impacts for a typical community.

Table 3.2: List of Impacts from Implementing Adaptation Strategies

Impact	Tool or Measurement Methodology	
	Non-Market Valuation Methodologies	Benefits Transfer
Benefits		
Recreation with enhanced water quality	•	•
Property value increase in a protected community	•	•
Enhanced ability to attract new business	•	•
Quality of life (decreased anxiety, increased safety)	•	•
Enhanced aesthetics	•	•
Other ecosystem service related benefits	•	•
Other Costs		
Decreased aesthetics	•	•
Decreased access to a beach or harbor	•	•
Environmental damage from constructing or implementing the resilient infrastructure	•	•
Other ecosystem service related costs	•	•

Task 2: Monetize Impacts

Objective: Select tools and approaches to monetize costs and benefits of the impacts identified in Task 1 of this chapter based on available resources.

Process to complete this task:

Step 1: Determine which tools and approaches you will use to monetize your list of impacts. Review Tables 3.1 and 3.2 from the previous task to see the type of impacts each tool or approach can monetize. Review **Appendix B,** which includes more detailed information about all of the tools and approaches, to learn more about the approach, including a description, general process for performing it, level of effort needed, expertise needed, and important resources. Keep in mind that Appendix B does not list all possible monetization approaches—consult your economic expert for more suggestions. As for many of the comprehensive tools and monetization approaches, you can monetize many primary impacts by answering three questions.

- How high is the water for each flooding scenario?

- How badly will the water damage your priority assets?

- How much is the asset worth?

While comprehensive tools often rely on more generic data, they can be quite useful when performing the holistic approach and monetizing damage on a large number of infrastructure. However, when you are monetizing damage for priority infrastructure, you should use as much site-specific information as possible.

Key considerations for this task:

- Will you be able to monetize the most significant impacts? These typically include damage to priority infrastructure, buildings, and building contents, and sometimes business or utility interruption.

- Does it make sense for your community to use a comprehensive monetization tool —such as COAST or FEMA Hazus-MH Flooding—when performing this task? These are resource intensive but are sometimes your best option when performing the holistic approach.

- Can you monetize the impacts on your priority infrastructure without using comprehensive tools? Priority infrastructure can be unique, and you will generate more accurate damage estimates by leveraging site-specific data than relying on comprehensive monetization tools that may rely on more generic data.

- Is it possible to quantify any of the impacts that you cannot monetize? Doing so will provide more insight than just considering qualitatively.

Experts required for this task:

- Experienced economists to help you monetize the costs and benefits in Steps 1, 2, 3, 4, and 6.

- GIS analysts to help your team use spatial data in comprehensive monetization tools in Step 4.

Step 2: List the tool or approach you will use to monetize or quantify each impact you listed in Task 1 of the chapter. Otherwise, note that you will only consider that impact qualitatively.

Step 3: From the list of primary, secondary, and environmental impacts you plan to monetize or quantify, create tables that show which impacts are applicable at each water-level increase for your no-action scenario as well as each action scenario. This will help you focus on the impacts you need to monetize or quantify at each water-level increase. The contents of these tables will depend somewhat on whether you are performing the impact or risk assessment. The impact assessment approach involves creating one table with a column for each action scenario, whereas the risk assessment approach involves creating a separate table for each action scenario, including the no action scenario for each SLR scenario selected, so that you can assess costs for the different storm types and years selected in Chapter 1. If you performed an impact assessment as part of Chapter 1 Task 2, continue to use Option A. If you elected to perform a risk assessment, continue to follow Option B of this task.

Option A: impact assessment

Create a table listing the impacts due to coastal flooding that you plan to quantify or monetize. List the primary, secondary, and environmental impacts that are applicable at each water-level increase for the no-action and each action scenario. Table 3.3 shows how this table can look, however you will input monetized cost data in the next step. Keep in mind that many impacts may only be applicable for higher water-level increases or in a no-action scenario. For example, an action scenario that implements engineered barriers may eliminate impacts at certain water-level increases. If you are following the priority infrastructure approach, you may want to provide more detailed site-specific information than is shown in Table 3.3 such as the specific content in the infrastructure damaged or components of the infrastructure that would be affected at each water-level increase.

Table 3.3: Sample Table of Monetized Damage for the Impact Assessment

Water-Level Increase	No-Action Scenario Damage ($million)*	Action Scenario 1 Damage ($million)	Action Scenario 2 Damage ($million)
3 feet	**PRIMARY DAMAGE** Residential: $50 Commercial: $30 Public Infrastructure: $15 **SECONDARY DAMAGE** Business interruption: $5	**PRIMARY DAMAGE** Residential: None Commercial: None Public Infrastructure: None **SECONDARY DAMAGE** Business interruption: None	**PRIMARY DAMAGE** Residential: $25 Commercial: $15 Public Infrastructure: $7 **SECONDARY DAMAGE** Business interruption: $3
8 feet	**PRIMARY DAMAGE** Residential: $500 Commercial: $300 Public Infrastructure: $150 **SECONDARY DAMAGE** Business interruption: $50	**PRIMARY DAMAGE** Residential: None Commercial: None Public Infrastructure: None **SECONDARY DAMAGE** Business interruption: None	**PRIMARY DAMAGE** Residential: $250 Commercial: $150 Public Infrastructure: $75 **SECONDARY DAMAGE** Business interruption: $25
12 feet	**PRIMARY DAMAGE** Residential: $2,000 Commercial: $1,200	**PRIMARY DAMAGE** Residential: None Commercial: None	**PRIMARY DAMAGE** Residential: $1,000 Commercial: $600

Table 3.3: Sample Table of Monetized Damage for the Impact Assessment

Water-Level Increase	No-Action Scenario Damage ($million)*	Action Scenario 1 Damage ($million)	Action Scenario 2 Damage ($million)
	Public Infrastructure: $600 **SECONDARY DAMAGE** Business interruption: $200	Public Infrastructure: None **SECONDARY DAMAGE** Business interruption: None	Public Infrastructure: $300 **SECONDARY DAMAGE** Business interruption: $100
18 feet	**PRIMARY DAMAGE** Residential: $8,000 Commercial: $4,800 Public Infrastructure: $2,400 **SECONDARY DAMAGE** Business interruption: $800	**PRIMARY DAMAGE** Residential: $8,000 Commercial: $4,800 Public Infrastructure: $2,400 **SECONDARY DAMAGE** Business interruption: $800	**PRIMARY DAMAGE** Residential: $4,000 Commercial: $2,400 Public Infrastructure: $1,200 **SECONDARY DAMAGE** Business interruption: $400

*This is a simplified table for illustrative purposes and does not include all impacts such as environmental impacts. It is recommended to monetize as many impacts in your table as possible from Tables 3.1 and 3.2.

Option B: risk assessment

Create tables, one for the no-action scenario and one for each action scenario for each SLR scenario selected, for the impacts due to coastal flooding that you plan to quantify or monetize. List the primary, secondary, and environmental impacts that are applicable to each storm type and year you selected in Task 2 of Chapter 1. At this point, you may want to modify the years if you choose to change the timeframe of your analysis. You will need a separate set of tables for each SLR scenario you selected. Table 3.4 shows how one table might look for a sample no-action scenario (though at this step you would not yet have included monetized cost data, which is added in the next steps). Keep in mind that many impacts may only be applicable at higher levels of inundation or in a no-action scenario. For example, an action scenario that implements engineered barriers may eliminate impacts at certain water-level increases. If you are following the priority infrastructure approach, you can provide more detailed site-specific information than is shown in Table 3.3. For example, consider including the specific content of the infrastructure damaged or components of the infrastructure that would be affected at each water-level increase.

Table 3.4: Sample Table of Monetized Damage for the Risk Assessment

Storm Type	2010 Damage ($million)*	2040 Damage ($million)	2070 Damage ($million)	2100 Damage ($million)
1-year	**PRIMARY DAMAGE** Residential: $75 Public Infrastructure: $20 **SECONDARY DAMAGE** Business interruption: $5	**PRIMARY DAMAGE** Residential: $90 Public Infrastructure: $25 **SECONDARY DAMAGE** Business interruption: $5	**PRIMARY DAMAGE** Residential: $120 Public Infrastructure: $22 **SECONDARY DAMAGE** Business interruption: $8	**PRIMARY DAMAGE** Residential: $150 Public Infrastructure: $40 **SECONDARY DAMAGE** Business interruption: $10

Table 3.4: Sample Table of Monetized Damage for the Risk Assessment

Storm Type	2010 Damage ($million)*	2040 Damage ($million)	2070 Damage ($million)	2100 Damage ($million)
2-year	**PRIMARY DAMAGE** Residential: $150 Public Infrastructure: $40 **SECONDARY DAMAGE** Business interruption: $10	**PRIMARY DAMAGE** Residential: $200 Public Infrastructure: $135 **SECONDARY DAMAGE** Business interruption: $15	**PRIMARY DAMAGE** Residential: $350 Public Infrastructure: $200 **SECONDARY DAMAGE** Business interruption: $50	**PRIMARY DAMAGE** Residential: $750 Public Infrastructure: $200 **SECONDARY DAMAGE** Business interruption: $50
10-year	**PRIMARY DAMAGE** Residential: $750 Public Infrastructure: $200 **SECONDARY DAMAGE** Business interruption: $50	**PRIMARY DAMAGE** Residential: $950 Public Infrastructure: $300 **SECONDARY DAMAGE** Business interruption: $50	**PRIMARY DAMAGE** Residential: $1,200 Public Infrastructure: $400 **SECONDARY DAMAGE** Business interruption: $100	**PRIMARY DAMAGE** Residential: $1,875 Public Infrastructure: $500 **SECONDARY DAMAGE** Business interruption: $125
30-year	**PRIMARY DAMAGE** Residential: $1,875 Public Infrastructure: $500 **SECONDARY DAMAGE** Business interruption: $125	**PRIMARY DAMAGE** Residential: $2,200 Public Infrastructure: $1,100 **SECONDARY DAMAGE** Business interruption: $200	**PRIMARY DAMAGE** Residential: $3,200 Public Infrastructure: $1,500 **SECONDARY DAMAGE** Business interruption: $300	**PRIMARY DAMAGE** Residential: $7,500 Public Infrastructure: $2,000 **SECONDARY DAMAGE** Business interruption: $500
100-year	**PRIMARY DAMAGE** Residential: $7,500 Public Infrastructure: $2,000 **SECONDARY DAMAGE** Business interruption: $500	**PRIMARY DAMAGE** Residential: $15,000 Public Infrastructure: $4,000 **SECONDARY DAMAGE** Business interruption: $1,000	**PRIMARY DAMAGE** Residential: $22,000 Public Infrastructure: $11,000 **SECONDARY DAMAGE** Business interruption: $2,000	**PRIMARY DAMAGE** Residential: $38,000 Public Infrastructure: $18,000 **SECONDARY DAMAGE** Business interruption: $4,000
200-year	**PRIMARY DAMAGE** Residential: $32,000 Public Infrastructure: $15,000 **SECONDARY DAMAGE** Business interruption: $3,000	**PRIMARY DAMAGE** Residential: $48,000 Public Infrastructure: $22,000 **SECONDARY DAMAGE** Business interruption: $6,000	**PRIMARY DAMAGE** Residential: $70,000 Public Infrastructure: $35,000 **SECONDARY DAMAGE** Business interruption: $10,000	**PRIMARY DAMAGE** Residential: $100,000 Public Infrastructure: $60,000 **SECONDARY DAMAGE** Business interruption: $12,000

*This is a simplified table for illustrative purposes and does not include all damage such as damage from environmental impacts. It is recommended to monetize as many impacts in your table as possible from Tables 3.1 and 3.2.

Step 4: Perform the monetization or quantification, and list the monetized or quantified value next to the impacts in the table(s) you created in the previous step of this process.

Step 5: Create a table similar to Table 3.5 or 3.6 that compiles the total of all your monetized impacts. This table's contents will depend on whether you are performing the impact or risk assessment.

Option A: impact assessment

Sum the total of all monetized damage for each water-level increase for the no-action scenario and each action scenario. Table 3.5 provides an example, showing how costs from Table 3.3 were compiled.

Table 3.5: Sample Table of Compiled Monetized Damage for the Impact Assessment

Water-Level Increase	No-Action Scenario Damage ($million)	Action Scenario 1 Damage ($million)	Action Scenario 2 Damage ($million)
3 feet	$100	$0	$50
8 feet	$1,000	$0	$500
12 feet	$4,000	$0	$2,000
18 feet	$16,000	$16,000	$8,000

Option B: risk assessment

Sum the total of all monetized damage for each storm type at each year. Create one table for each table developed in the previous step of this process; include a column for the annual probability of each storm type. For example, the 2-year storm occurs on average once every 2 years, so its annual probability is 0.5. Table 3.6 provides an example, showing how costs from Table 3.4 were compiled for a sample no-action scenario.

Table 3.6: Sample Table of Compiled Monetized Damage for the Risk Assessment

Storm Type	Annual Probability	2010 Damage ($million)	2040 Damage ($million)	2070 Damage ($million)	2100 Damage ($million)
1-year	1	$100	$120	$150	$200
2-year	0.5	$200	$350	$600	$1,000
10-year	0.1	$1,000	$1,300	$1,750	$2,500
30-year	0.033	$2,500	$3,500	$5,500	$10,000
100-year	0.01	$10,000	$20,000	$35,000	$60,000
200-year	0.005	$50,000	$76,000	$115,000	$172,000

Step 6: Monetize or quantify all of the benefits and other costs that you can from the list you created in Task 1, Step 2 of this chapter. These are typically independent of the water-level increase. You will probably just need to do this once for each benefit or other cost within each action scenario.

Task 3: Estimate Costs of Implementing Adaptation Strategies

Objective: Identify and estimate all capital and maintenance costs of implementing the adaptation strategies in your action scenario.

Process to complete this task:

Step 1: Get existing cost data about similar projects in your community or other communities. You may be able to find this information by reviewing **Appendix A** for your adaptation strategy, performing Internet searches for case studies, or asking other community leaders about their projects.

Key considerations for this task:

- Can you find cost data from other similar adaptation projects? This can be an easy way to do a preliminary analysis of your best action scenarios. If you decide that an action scenario is a top candidate to implement, consider getting a more precise estimate from an architecture or engineering firm.

Experts required for this task:

- Engineers to help you estimate the costs of adaptation strategies in Step 2.

Step 2: Develop estimates for the capital costs, maintenance costs, and timing of each maintenance cost for each adaptation strategy within each action scenario using existing data and parametric software, or by consulting an architectural or engineering firm. Keep in mind that some capital costs will be incurred in the near future and others will not be incurred until later in the planning horizon. Try to get specification details and information about the useful life of each measure. Do not overlook planning costs in the initial capital costs or administrative costs in the maintenance costs.

Table 3.7: Parametric Software

Software	Overview
RS Means	Used for high-level "assembly" costs, as well as more detailed estimates for individual building systems and components. *Level of technical expertise needed:* An engineering background and training with this software. *Availability:* Check with your local planning public works engineering offices.* *Cost:* $100 to $1,000.
RS Means Quick Cost Estimator	Online database providing planning-level costs based on building type, square footage, and location; however, it may only apply to limited building-related adaptation strategies. *Level of technical expertise needed:* It is a basic online system that requires simple entry of building type, square footage, and location. *Availability:* Online. *Cost:* Free.

Table 3.7: Parametric Software

Software	Overview
U.S. Cost Success Estimator	Performs detailed bottom-up analyses. Provides the ability to integrate with commonly used design tools and draw from multiple cost databases. ***Level of technical expertise needed:*** An engineering background and training with this software. ***Availability:*** Check with your local planning or public works engineering offices. ***Cost:*** $4,295.
Parametric Cost Engineering System	Prepares parametric cost estimates for new facility construction, renovation, and life cycle cost analysis using pre-engineered model parameters and construction criteria. ***Level of technical expertise needed:*** An engineering background and training with this software. ***Availability:*** Check with your local planning or public works engineering offices. ***Cost:*** $850 for federal government users; $3,000 all other users.

* If your local community does not have a public works engineering office or another local planning office with someone trained to use one of these parametric software programs, it can be quite resource intensive to invest in training, and it may be worthwhile to consult an engineering firm.

Chapter 4: Make a Decision

Now you will compile the monetized, quantitative, and qualitative data you developed in the previous chapter and decide whether your action scenario makes financial sense—whether the benefits outweigh the costs. If you prepared multiple action scenarios, more than one of them might make financial sense; in that case, you will want to find the best option for your community based on the cost-benefit results, financial feasibility, and other relevant considerations.

Case Study: Old Orchard Beach, Maine

Old Orchard Beach, a popular summer beach destination in southern Maine, assessed three action scenarios for dealing with increased coastal flooding due to SLR: taking no action, nourishing the beach within the 100-year floodplain, and nourishing the beach within the 50-year floodplain. The city reviewed each scenario under high, low, and no SLR conditions, concluding that the robust decision was to nourish the beach within the 100-year floodplain. With no SLR, the city predicted that this action would have only slightly higher costs than the 50-year nourishment strategy; with low or high SLR, they predicted lower damage and costs. They also found that taking no action would result in higher costs under all SLR conditions.

Task 1:
Calculate Total Benefits of Each Action Scenario

Task 2:
Compile the Capital and Maintenance Costs

Task 3:
Assess Each Action Scenario

Outputs:
- Total Benefits of Action Scenarios
- Timelines of Capital and Maintenance Costs
- Ranking of Action Scenarios

Expertise required to complete Chapter 4:
- Experienced **economists** to help you determine a discount rate in Task 2 and interpret your results in Task 3

Key resources referenced in Chapter 4 include:
- Discount rate guidance: Office of Management and Budget's Circular A-4, "Regulatory Analysis"
- Discount rate guidance: EPA's *Guidelines for Preparing Economic Analyses*, Chapter 6

Task 1: Calculate Total Benefits of Each Action Scenario

Objective: Calculate the total benefits of each action scenario for each high-water-level event using damage and other costs and benefits monetized in Chapter 3.

Process to complete this task:

The steps in this process are different for the impact and risk assessment. If you performed an impact assessment as part of Chapter 1 Task 2, continue to use Option A. If you performed a risk assessment, continue to follow Option B of this task.

> **Key considerations for this task:**
>
> • What discount rate will you choose if you are performing the risk assessment?
>
> • Would it be helpful to your discussion to choose multiple discount rates?

Option A: impact assessment

Step 1: Create a table that lists total benefits by water-level increase for each action scenario, similar to the example shown in Table 4.1, which shows the total benefits of just one action scenario. The rows for this table will correspond to the water-level increases you chose to assess in Chapter 1. Steps 2 to 4 below explain more about where to gather data and how to make the calculations for this table.

Table 4.1: Example Table of Total Benefits for the Impact Assessment

Water-Level Increase	No-Action Scenario Damage ($million)	Action Scenario 1 Damage ($million)	Action Scenario 1 Other Monetized Benefits ($million)	Action Scenario 1 Other Costs ($million)	Action Scenario 1 Total Monetized Benefits ($million)
3 feet	$100	$50	$25	$5	$70
8 feet	$1,000	$500	$25	$5	$520
12 feet	$4,000	$2,000	$25	$5	$2,020
18 feet	$16,000	$8,000	$25	$5	$8,020

Step 2: Add monetized damage of inundation for no-action and action scenario(s) from Step 5 of Task 2 in Chapter 3 to this table.

Step 3: Add other costs and benefits data for each action scenario from Step 6 of Task 2 in Chapter 3 to this table. These other costs and benefits are typically the same across water-level increases. As discussed in Chapter 3, they can often be quite difficult to assign a dollar value to—you may need to just consider them qualitatively, in which case you would put $0 into the appropriate fields of this table but footnote the qualitative costs and benefits.

Step 4: Calculate the total benefits for each water-level increase using Equation 4.1 and add to your table:

$$Ben_T = IC_{NA} - IC_A + Ben_A - OC_A \qquad \text{(Equation 4.1)}$$

$$\$8,020 \text{ (million)} = \$16,000 - \$8,000 + \$25 - \$5$$
(sample data from Table 4.1 at an 18-foot water-level increase)

Where:

Ben_T	=	total benefits
IC_{NA}	=	no-action scenario damage
IC_A	=	action scenario damage
Ben_A	=	action scenario benefits
OC_A	=	other monetized action scenario other costs

In many cases, your benefits and other costs may just be qualitative or quantitative, and your monetized total benefits will simply be calculated from avoided costs: the difference in the losses from flooding associated with no-action and those associated with your action scenario. Non-monetized benefits should be described and, to whatever extent possible, quantified. These benefits should be considered in the final plan selection, as shown in Table 4.7 below.

Qualitative and Quantitative Benefits

Qualitative benefits are simply a description of the benefit. These should always be considered in decision making as they can be substantial, but it can often be difficult to weigh them against Monetized costs, which have been assigned dollar value. Quantitative benefits provide additional information by assigning a non-monetary value to the benefit such as number of fishing days enhanced by improved water quality. When it is not possible to monetize benefits, you should at least try to quantify them as it provides further information beyond a qualitative description for making informed cost-benefit decisions.

Option B: risk assessment

Step 1: Create tables of the average and expected damage for select years, similar to the example shown in Table 4.2, which shows damage for the sample no-action scenario. The equations in subsequent steps show how to calculate the values for the tables. Average damage provides an average across storms of varying severity, and expected damages factors in the probability of those different storms occurring. Create one table under each SLR scenario for the no-action scenario and each action scenario. Calculations for this table are shown in Steps 2 to 4 below.

Table 4.2: Sample Table of Annual Loss for Select Years for the Risk Assessment

Year		2010		2040		2070		2100	
Storm Type	Annual Probability	Average Damage ($billion)	Expected Damage ($billion)	Average Damage ($billion)	Expected Damage ($billion)	Average Damage ($billion)	Expected Damage ($billion)	Average Damage ($billion)	Expected Damage ($billion)
1- to 2-year	0.5	0.15	0.08	0.24	0.12	0.38	0.19	0.60	0.30
2- to 10-year	0.4	0.60	0.24	0.83	0.33	1.18	0.47	1.75	0.70
10- to 30-year	0.067	1.75	0.12	2.40	0.16	3.63	0.24	6.25	0.42
30- to 100-year	0.023	6.25	0.15	11.75	0.27	20.25	0.47	35.00	0.82
100- to 200-year	0.005	35.00	0.15	48.00	0.24	75.00	0.38	116.00	0.58
Beyond 200-year	0.005	50.00	0.25	76.00	0.38	115.00	0.58	172.00	0.86
Expected Annual Loss ($billion)			0.98		1.50		2.32		3.67

Step 2: Calculate the annual probability of consecutive storm types for each year in your table using Equation 4.2, and add these data to your table. For the biggest storm type, the annual probability is just the probability of that storm type. This is the second column of Table 4.2.

$$AP_{XY} = AP_X - AP_Y \qquad \text{(Equation 4.2)}$$

$$0.4 = 0.5 - 0.1$$
(sample data from Table 4.2 for the annual probability of a 2- to 10-year storm)

Where:

X and Y	=	different storm types, for example, a 1-year and 2-year storm
AP_{XY}	=	annual probability of storm type X to storm type Y
AP_X	=	annual probability of storm type X
AP_Y	=	annual probability of storm type Y

Step 3: Calculate the average damage of consecutive storm types for each year in your table using Equation 4.3, and add these data to your table. For the biggest storm type, the average damage is just the monetized damage for that storm type from Step 5 of Task 2 in Chapter 3. These data are calculated for four different years in Table 4.2.

$$AD_{XYN} = (MD_{XN} + MD_{YN}) \div 2 \qquad \text{(Equation 4.3)}$$

$$\$0.83\ (billion) = (\$0.35 + \$1.30) \div 2$$
(sample data from Table 3.6 for 2- to 10-year storm in 2040)

Where:

AD_{XYN}	=	average damage of storm type X to storm type Y at year N
N	=	year
MD_{XN}	=	monetized damage of storm type X at year N
MD_{YN}	=	monetized damage of storm type Y at year N

Step 4: Calculate the expected damage of consecutive storm types for each year in your table using Equation 4.4, and add these data to your table. These data are calculated for four different years in Table 4.2.

$$ED_{XYN} = AD_{XYN} \times AP_{XY} \qquad \textbf{(Equation 4.4)}$$

$$\$0.33\ (billion) = \$0.83 \times 0.4$$
(sample data from Table 4.2 for 2- to 10-year storm in 2040)

Where:

ED_{XYN}	=	expected damage of storm type X to storm type Y at year N
AP_{XY}	=	annual probability of storm type X to storm type Y
AD_{XYN}	=	average damage of storm type X to storm type Y at year N

Step 5: Calculate the expected annual loss across all storm types for each year by summing the expected damage for all storm types in your table for that given year. This is shown in the bottom row of Table 4.2.

Step 6: Create timelines showing the expected annual loss for each year, as shown in Table 4.3, which shows a timeline for the sample no-action scenario. Create one timeline for each table developed in Step 2 of this process. Create columns at annual increments, beginning with the first year for which you calculated losses and ending at the last year of your assessment. Steps 7 and 8 explain more about completing this timeline.

Table 4.3: Sample Timeline of Annual Loss for the Risk Assessment

Year	2010	2011	2039	2040	2069	2070	2099	2100
Expected Annual Loss ($billion)	0.98	0.99	1.48	1.50	2.29	2.32	3.63	3.67

Note: The years in this table should continue uninterrupted from the first year of the analysis until the final year.

Step 7: Input the expected annual losses from Step 5 directly into your timeline. In the example provided in Table 4.2, this includes data from 2010, 2040, 2070, and 2100.

Step 8: Calculate the values for each year between the selected years from Step 6. Assume that the losses increase or decrease linearly between select years. For example, in Table 4.3 the damage is $1.50 billion in 2040 and $2.32 billion in 2070, and the average annual increase over that 30-year window is $0.027 billion. Thus, the annual loss is assumed to increase by $0.027 billion starting in 2040 until 2070.

Step 9: If the benefits or other costs are monetized and differ from select year to select year, in the absence of a methodology that better estimates the monetized values for the years between your estimates, repeat Steps 7 and 8 above to create timelines for the value of benefits or other costs.

Step 10: Compile data from your timelines in the preceding steps to create a comprehensive timeline (as illustrated in Table 4.4 for one action scenario) showing annual total benefits, present value of annual total benefits, and net present value (NPV) of total benefits. Create one timeline for each action scenario under each SLR scenario. You can choose to include one or multiple discount rates. Steps 11 to 13 provide more information about where to get data and how to make calculations for this table.

Table 4.4: Sample Timeline of the Present Value of Total Benefits for the Risk Assessment

Year	2010	2011	2039	2040	2069	2070	2099	2100	
Expected Annual Loss (No Action) ($billion)	0.98	0.99	1.48	1.50	2.29	2.32	3.63	3.67	
Expected Annual Loss (Action Scenario 1) ($billion)	0.33	0.33	0.49	0.50	0.76	0.77	1.21	1.22	
Action Scenario Annual Benefits ($billion)	0.05	0.05	0.05	0.05	0.05	0.05	0.05	0.05	
Action Scenario Other Costs ($billion)	0.02	0.02	0.02	0.02	0.02	0.02	0.02	0.02	
Annual Total Benefits ($billion)	0.68	0.69	1.02	1.03	1.56	1.58	2.45	2.48	**NPV**
Present Value of Total Benefits (3 Percent Discount Rate) ($billion)	0.68	0.67	0.43	0.42	0.27	0.27	0.18	0.17	33.95
Present Value of Total Benefits (No Discounting) ($billion)	0.68	0.69	1.02	1.03	1.56	1.58	2.45	2.48	127.26

Note: The years in this table should continue uninterrupted from the first year of the analysis until the final year.

Step 11: Pull expected annual loss data for both your no-action and action scenario created in Steps 6 through 8, as well as annual benefits and other cost data created in Step 9, into your comprehensive timeline. Refer to the first four rows of Table 4.4.

Step 12: Calculate annual total benefits for each year using Equation 4.5, and add these data to your table. In Table 4.4, these data correspond to the fifth row below the header.

$$Ben_{TN} = IC_{NA-N} - IC_{AN} + Ben_{AN} - OC_{AN}$$ (Equation 4.5)

$0.69 (billion) = $0.99 - $0.33 + $0.05 - $0.02
(sample data from Table 4.4 *for 2011* total benefits)

Where:

Ben_{TN}	=	total benefits at year N
IC_{NA-N}	=	no-action scenario damage at year N
IC_{AN}	=	action scenario damage at year N
Ben_{AN}	=	action scenario benefits at year N
OC_{AN}	=	action scenario other costs at year N

Step 13: Calculate the present value of total benefits for each year using Equation 4.6, and add these data to your table. You can make this calculation for one or more discount rates. In Table 4.4, these data correspond to the bottom two rows.

$$Ben_{PVND} = Ben_{TN} / (1 + D)^{(N-Y0)}$$ (Equation 4.6)

$0.27 (billion) = $1.58 / (1 + 0.03)^{(2070-2010)}
(sample data from Table 4.4 *for 2070* net present value of total benefits)

Where:

Ben_{PVND}	=	present value of benefits at year N for discount rate D
Ben_{TN}	=	total benefits at year N
D	=	discount rate
N	=	year you are assessing in the table
$Y0$	=	current year

Discounting Future Dollars—The "Discount Rate"

- Dollar values associated with future costs and benefits are typically discounted, because a dollar received today is considered more valuable than one received in the future. This allows dollar values associated with long-term costs and benefits to be compared equally against the values of short-term costs and benefits.

- The discount rate is the rate at which society as a whole is willing to trade present for future benefits, and it helps to account for inflation and other factors. However, determining the discount rate can be a source of controversy, and as shown in Table 4.4, it can significantly affect the analysis. Therefore, it would be helpful to consult an economic expert before selecting a discount rate.

Step 14: Calculate NPV of total benefits at each discount rate chosen. See Table 4.6 for resources about choosing a discount rate. The NPV is simply the sum of the present value of benefits for a given discount rate across all years. In Table 4.4, these data correspond to the bottom two rows in the last column.

Task 2: Calculate the Capital and Maintenance Costs

Objective: Determine the NPV of total costs for each action scenario using cost data from Task 3 of Chapter 3.

Process to complete this task:

Step 1: Create a timeline outlining the capital and maintenance costs using data outputs from Task 3 in Chapter 3 for each adaptation strategy within each of your action scenarios. Table 4.5 shows what this timeline

Key considerations for this task:
- What discount rate will you choose if you are performing the risk assessment?
- Would it be helpful to your discussion to choose multiple discount rates?

Experts required for this task:
- Economists to help you determine a discount rate in Step 4.

might look like for a given action scenario. Create one table for each action scenario. Add a row for each adaptation strategy within an action scenario, as well as rows for total costs, present value of total costs, and NPV of total costs. The column headers will include capital costs and maintenance costs for any year in which you expect maintenance to be performed for any adaptation strategy. Steps 2 to 6 provide further information about where to get data and how to make calculations for this table.

Table 4.5: Sample Timeline of Capital and Maintenance Costs and NPV

Cost Type	Capital ($million)	Maintenance ($million)	Maintenance ($million)	Maintenance ($million)	Maintenance ($million)	Maintenance ($million)
Year	2010	2020	2030	2040	2050	2060
Adaptation Strategy 1	$180	$175	$175	$175	$175	$175
Adaptation Strategy 2	$3,000	$17	$17	$17	$17	$17
Adaptation Strategy 3	$7,000	$0	$0	$0	$0	$0
Total cost for Action Scenario 1	$10,180	$192	$192	$192	$192	$192
Present value of total costs*	$10,180	$143	$106	$79	$59	$44
NPV of Total Costs ($million)	$10,611					

*   Assumes a 3 percent discount rate.

Step 2: Add capital and maintenance cost information to this table from your outputs in Task 3 in Chapter 3.

Step 3: Calculate the total capital and maintenance cost for each year in your table by summing these costs across all adaptation strategies within a given year.

Step 4: Choose a discount rate or multiple discount rates in order to calculate the present value of these costs. Consult your economic expert and Table 4.6 below.

Table 4.6: Resources for Choosing a Discount Rate

Resource	Key Information
State agencies	Some states have guidance on using discount rates and may have adopted rates to use for performing cost-benefit analyses.
Office of Management and Budget's Circular A-4, "Regulatory Analysis"	Provides information on using discount rates in cost-benefit analyses, including rationale for using discount rates, considerations about intergenerational discounting, and a recommendation to at least consider 3 percent and 7 percent discount rates.
EPA's *Guidelines for Preparing Economic Analyses,* **Chapter 6**	Provides information on using discount rates in cost-benefit analyses, including an overview of discounting, information about NPV, issues in applying discount rates, and recommended approaches.

Step 5: Calculate the present value of capital and maintenance costs at each year using Equation 4.7, and add these data to your table. This equation applies to both future maintenance and future capital costs. You can perform this calculation for one or more discount rates. In Table 4.4, these data correspond to the bottom two rows.

$$Cost_{PVND} = Cost_{TN} / (1 + D)^{(N-Y0)}$$ (Equation 4.7)

$$\$106 \ (million) = (\$175 + \$17 + \$0) / (1 + 0.03)^{(2030-2010)}$$
(sample data from Table 4.5 for maintenance costs in 2030)

Where:

$COST_{PVND}$	=	present value of costs at year N for discount rate D
$Cost_{TN}$	=	total costs at year N (from step 3)
D	=	discount rate
N	=	year you are assessing in the table
$Y0$	=	current year

Step 6: Calculate NPV of total capital and maintenance costs at each discount rate selected. This is simply the sum of the present value of costs for a given discount rate across all years. In Table 4.5, this value is in the bottom row.

Task 3: Assess Each Action Scenario

Objective: Identify the action scenarios, if any, whose total benefits exceed total costs—both monetized and qualitative. Rank these based on the cost-benefit results, financial feasibility, and any other considerations you might have.

Process to complete this task:

Step 1: Assess the monetized cost-benefit results of each action scenario. This step differs slightly for the risk and impact assessments.

Option A: impact assessment

Compare the NPV of total costs calculated in Task 2 of this chapter to the total benefits of each action scenario calculated in Task 1. In determining whether an action scenario makes financial sense, consider what types of coastal flooding events must occur for total benefits to outweigh costs. One or two frequent, low-water-level increase scenarios might justify an action, perhaps indicating a worthwhile scenario. Alternatively, it might require 10 huge high-water-level events for total benefits to exceed NPV of total costs, perhaps indicating a bad financial decision.

Option B: risk assessment

Calculate the net benefits—the difference in NPV of total benefits calculated in Task 1 of this chapter and NPV of total costs calculated in Task 2 of this chapter. Calculate the benefit-to-cost ratio—NPV of total benefits divided by NPV of total costs. Make these calculations for each action scenario under each SLR scenario. Table 4.7 provides an example.

Key considerations for this task:

- Did you factor qualitative and quantitative impacts and benefits into your monetized results? These can help you choose the best action scenario or decide whether an action scenario is worthwhile, especially when the monetized results are less conclusive.

- Did you assess the feasibility of paying both capital and maintenance costs in each action scenario? Communities often neglect to plan how to finance maintenance and administrative costs.

Experts required for this task:

- Economists to help you interpret your results in Steps 1 and 2.

Table 4.7: Sample Net Benefit and Benefit to Cost Calculations for a Risk Assessment

	NPV of Total Benefits*	NPV of Total Costs	Net Benefits	Benefit-to-Cost Ratio	Non-Monetized Benefits
Action Scenario 1	$33.95 billion	$10.61 billion	$23.34 billion	3.2:1	Medium
Action Scenario 2	$1.5 billion	$0.5 billion	$1 billion	3:1	High
Action Scenario 3	$25 million	$5 million	$20 million	5:1	Low

* The NPV of Total Benefits and NPV of Total Costs for Action Scenario 1 are pulled from Tables 4.4 and 4.5, respectively, and assume a 3 percent discount rate. Values for Action Scenario 2 and 3 were not derived from another table in this framework but are included for illustrative purposes.

Step 2: For each action scenario, decide whether the benefits outweigh the costs. As well as the monetized costs and benefits you have just tabulated, consider qualitative and quantitative impacts and benefits. When the monetized results are less conclusive, these considerations may make the difference. (When the monetized results are overwhelming in one direction, on the other hand, they may not affect the outcome.) Also consider the implications of non-monetized co-benefits described in Task 1 of this Chapter. These non-monetized benefits can tip the balance in favor of scenarios with a lower monetized benefit-to-cost ratio, indicating that they may warrant a higher ranking than would be indicated by monetized values alone.

Step 3: Assess the feasibility of funding each action scenario that you found to be potentially worthwhile. Determine the feasibility of funding the capital and maintenance costs for each action scenario.

Step 4: Rank your action scenarios from best to worst based on the first three steps of this process and any other factors you want to bring into the discussion.[4] Remember to factor in the outcome of the monetized cost-benefit analysis, qualitative and quantitative impacts, and funding feasibility. In the sample data in Table 4.7, for example, action scenario 1 could have the highest net benefits and the second highest benefit-to-cost ratio; however, for many communities, the inability to finance such a measure could make it an inferior option.

---

[4] The FEMA "STAPLEE" method is one process for considering a range of factors. This method is designed to help planning teams consider in a systematic way the Social, Technical, Administrative, Political, Legal, Economic, and Environmental (STAPLEE) opportunities and constraints of implementing a hazard mitigation action. This method is detailed in a FEMA guide entitled Developing the Mitigation Plan: Identifying Mitigation Actions and Implementation Strategies, which is available online at http://www.fema.gov/library/viewRecord.do?id=1886.

Conclusion: Economics is Only One Piece of Informed Decision-making

Understanding short and long-term costs and benefits of different adaptation strategies, as well as the costs of not taking action, is critical to making resilience-minded decisions that are fiscally and socially responsible. But community decisions are not based solely on economics. Other factors will need to be considered prior to implementing an action scenario—including social feasibility, community culture, and administrative and legal aspects. While economics are a very important criterion to consider, the unique conditions, history, and desired vision of each community will influence decisions about how to plan for and adapt to inundation threats from sea level rise and storms. As mentioned at the beginning of this guide, we recommend community leaders engage area stakeholders and residents in the decision process to develop a complete picture of community interests. Applying this framework can produce economic information that will be an important part of a community discussion.

Appendix A: Adaptation Strategies

The fact sheets in this appendix provide general, introductory information about a variety of adaptation strategies that communities may consider. Many of these strategies could be used to protect either private property or public infrastructure. Some of the strategies are traditionally employed to protect private property but could also be viable options to protect public property as well. This appendix is not intended to be a comprehensive inventory, but rather to introduce a range of strategies that help illustrate varying approaches: managed retreat policies, tidal management, engineered barriers, infrastructure modification/design, land use policy, and green infrastructure.

The fact sheets provide an overview of costs, effectiveness, and barriers to implementation. When reviewing these strategies for use in your own community, keep in mind that:

- Costs can vary significantly depending on the unique characteristics of your community and the exact nature of a project or policy,
- Non-economic factors such as legal challenges or public outreach needs can increase the resources needed to implement a strategy,
- The lifespan and effectiveness of any project will depend on the severity of future events.

Some fact sheets include special considerations for certain adaptation strategies, but when investigating any potential strategy a community will need to consider a number of factors such as social feasibility, environmental impacts, and administrative and legal aspects.

The current state of knowledge pertaining to adaptation strategies is growing rapidly as more and more communities seek and test out solutions to current and future inundation hazards. Communities should seek out additional sources of information to learn more about the strategies contained in this appendix and to identify additional strategies. The resources listed in the "Key References" text box in Chapter 2 of this framework as well as those that follow are a few potential sources of additional information:

- The Federal Emergency Management Agency (FEMA) has a variety of publications related to flood hazard mitigation measures, including guidance documents as well as case studies. FEMA recently released Mitigation Ideas: A Resource for Reducing Risk to Natural Hazards, which includes actions to mitigation risks from storm surge, flood, and sea level rise.
- The U.S. Army Corps of Engineers (USACE) has resources related to both structural and non-structural flood mitigation measures. Visit the Flood Risk Management for State and Local Partners section of the USACE Flood Risk Management Program website for more information.
- The Georgetown Climate Center has a searchable Adaptation Clearinghouse that provides more detailed information on a wide range of adaptation measures.
- NOAA's Adapting to Climate Change: A Planning Guide for State Coastal Managers includes a chapter on adaptation strategies, many of which are related to inundation hazards.

Fact Sheet A-1: Transfer of Development Rights

A transfer of development rights (TDR) ordinance provides a way for property owners to transfer development rights from one area to another. In the context of coastal flooding, this can be used as a measure to move future development from vulnerable areas to those that are completely out of harm's way. Additionally, TDR programs can be used to preserve open space, thereby facilitating the implementation of other mitigation measures, such as wetlands development or other green infrastructure to further increase a community's resilience to coastal flooding. TDRs are usually administered through a local government zoning ordinance, with specific districts zoned as either sending or receiving parcels. It is the local government's responsibility to determine the specific number and type of development rights (usually in terms of dwelling units or floor area per acre) that will be transferred. Once the transaction is complete, development rights are legally separated from the first "sending" parcel of land and attached to the second "receiving" parcel for use by the owner for development.[1]

Cost

TDR is a low-cost option for local governments because they are typically only responsible for the costs of developing and implementing the program through its planning and zoning functions, while the owners and sellers typically pay the costs for the development rights. Sometimes, local governments include transaction fees with the cost of the development rights when they are sold to help offset the costs of administering the program.

Effectiveness

TDR can be a highly effective and long-term solution to mitigate impacts of coastal flooding, as long as the development parcels are moved out of harm's way—to higher, further inland, and more protected areas.

Barriers to Implementation

TDR can be legally complex to implement and is generally not quickly implemented. For a program to be effective, TDR requires purchaser demand for the development rights as well as willing sellers.

Special Considerations

In some cases, a buyer, such as the local government or a nonprofit organization, may purchase the development rights to hold or sell them at a later date. In this case, the buyer could choose to permanently retire the development rights for conservation purposes

---

[1] American Planning Association (2006). *Planning and Urban Design Standards.*

Fact Sheet A-2: Purchase of Development Rights

Purchase of development rights (PDR) involves a local government or nonprofit purchasing development rights while the land remains privately owned. This restricts the future use of a property from certain types of development and is often used to preserve open space or farmland. In the context of coastal flooding, this can be used as a measure to prevent future development from occurring in vulnerable areas. PDRs are usually administered through a local government and often performed in collaboration with a nonprofit or trust that manages preservation activities.

Cost

PDR is a mid-price option with a higher cost than TDR but lower cost than a fee-simple acquisition—or outright buyout of the land. Costs for local government include those for developing and implementing the program as well as for purchasing the development rights. Purchase price for the development rights is estimated based on the most valuable use for the land allowed by zoning. For example, if the land is currently used for farming but could be developed for commercial uses based on current zoning laws, then the purchase price is generally based on the commercial value of the land.[2]

Effectiveness

Much like TDR, PDR is an effective and long-term solution; however, because participation in a PDR program is voluntary for the land owner, the programs can take a long time to implement over a large area. The PDR programs that would be most effective for limiting damage from coastal flooding are those that create large and contiguous areas with permanently preserved open space to serve as a buffer between flooding and development.

Barriers to Implementation

Because PDR programs rely on the voluntary participation of land owners, they generally are not quickly implemented. In addition, despite costing less than fee-simple acquisition, development rights can be prohibitively expensive for some local governments and are often tax-funded, which subjects PDR programs to political and economic risk.

Special Considerations

PDR usually only limits future development of housing, commercial, and similar functions; it does not remove existing development on a piece of land and does not prevent all future use of that land. The agreements generally allow the current land owner to stay on the land and, unless explicitly stated, do not prevent other activities, such as mining and farming.

---

[2] http://ohioline.osu.edu/cd-fact/1263.html

Fact Sheet A-3: Rolling Easements

Rolling easements are enforceable legal agreements that prohibit engineered barriers or other types of coastal armoring and that require removal of structures seaward of a migrating shoreline.[3] Rolling easements ensure that the shoreline will move naturally, and the shoreline is specifically defined (e.g. mean high water, vegetation line, or the upper boundary of tidal wetlands.) As the shoreline erodes or is inundated by sea level rise, existing development that ends up seaward of the defined line must be removed from harm's way. The easements may also potentially discourage future development in these areas. As the shoreline continues to recede, the easement "rolls" farther inland. The intent of rolling easements is to allow natural erosion to take place, which preserves natural sediment transport systems, wetlands, and other tidal habitats.[4] Easements can be purchased by a local government, donated by the land owner, or carried out through zoning.

Cost

Rolling easements are typically cheaper than protecting the land. In less developed areas, land owners are more likely to donate easements or sell them for a low price. Rolling easements for more densely developed areas have a higher cost because the land is more valuable and structures will have to be moved or abandoned as the shoreline moves in.

Effectiveness

When rolling easements lead to the removal of existing structures, they can prevent future flood damages in structure no longer subject to repeated or sustained inundation, but the value of the structure is also lost. Rolling easements are, however, a long-term strategy for potentially discouraging new housing developments along shorelines because they create the expectation that shoreline development will not be protected from flooding and erosion with engineered barriers or other types of coastal armoring. Thus, rolling easements have the potential long-term effect of mitigating future flood-related damage. Rolling easements can be most effective when implemented along undeveloped stretches of shoreline.

Barriers to Implementation

One method for implementing rolling easements is by obtaining voluntary participation of land owners; this can make the easements difficult to quickly implement, however, because property values are negatively impacted and existing structures cannot be protected with coastal armoring. Additionally, regulatory approaches toward implementation can often be legally complex.

Special Considerations

Rolling easements have sometimes been relaxed or removed in future years if the public starts becoming more sympathetic with regard to the destruction of waterfront housing.

---

[3] http://water.epa.gov/type/oceb/cre/upload/rollingeasementsprimer.pdf
[4] http://coastalmanagement.noaa.gov/initiatives/shoreline_ppr_easements.html

Fact Sheet A-4: Fee-Simple Acquisition

Fee-simple acquisition involves the outright purchase of property and all associated development rights. Fee-simple acquisition is often used when local governments purchase waterfront properties that are vulnerable to erosion and flooding.[5] In the context of coastal flooding, the purpose of the acquisition is to remove or prevent future development in vulnerable areas and to reduce future damage from coastal flooding. Fee-simple acquisitions can be used in conjunction with other managed retreat policies to preserve open space, which in turn can be used to implement other mitigation measures, such as wetlands development or green infrastructure, to further increase a community's resilience to coastal flooding.

Cost

Fee-simple acquisition is a high-cost option. The costs include purchasing the property and structures at fair value from a private owner. These costs are generally paid by local governments and financed by bonds (which can be difficult to implement) or federal or state grants (e.g., FEMA's Hazard Mitigation Grant Program.) Grants are sometimes available to help counties acquire important coastal lands. Fee-simple acquisition could also reduce property tax generation that had been collected from the purchased structures and/or land.

Effectiveness

Fee simple acquisition is highly effective because the public owns all rights to the land and can restrict development for as long as it maintains ownership.

Barriers to Implementation

The mechanisms for executing and financing fee-simple acquisition can be challenging. If property owners are unwilling to sell, the local government must exercise eminent domain in order to execute the purchase. Additionally, large-scale purchases can be prohibitively expensive for local governments to finance, although grants are sometimes available to help counties acquire important coastal lands.

Special Considerations

If local governments choose to pursue this through a forced buyout—or eminent domain—the process, cost, and ability of local governments can vary significantly depending on location.

---

[5] http://nationalshorelinemanagement.us/docs/Econ_Workshop_IWR04-NSMS-2.pdf

Fact Sheet A-5: Infrastructure Relocation

Infrastructure relocation involves moving vulnerable infrastructure to areas less susceptible to coastal flooding. Relocation can be a viable option for many types of infrastructure, including roads, bridges, buildings, utilities, and wastewater treatment plants. Moving infrastructure may involve physically relocating the existing infrastructure, constructing new replacement infrastructure, or otherwise shifting the function of the infrastructure to a different location. One way to shift the function to a less vulnerable area is through regionalization (e.g., expanding capacity at an existing waste water treatment plant in a less vulnerable area to eventually replace another wastewater treatment plant in a vulnerable area).

Cost

Cost varies depending on the infrastructure being relocated and can include the cost for new construction, moving lines, or physical relocation, as well as the cost of the land to which the infrastructure is being moved. Large and expensive infrastructure, such as power plants, sewer systems, and water treatment plants, can be very expensive to relocate, while roads and simple buildings are more financially feasible. Regionalization offers a lower cost alternative by having large infrastructure from a neighboring community replace the at-risk infrastructure in your community. With this approach, existing equipment is moved into the new facility, potentially increasing the capacity of the neighboring infrastructure. This could be a more cost-effective strategy for a wastewater treatment plant compared to relocating it because the land and supporting infrastructure (e.g., feeder lines) are already in place.

Effectiveness

Relocation is effective if the infrastructure is moved out of the vulnerable area. If the infrastructure is not moved completely outside the vulnerable area, the overall effectiveness depends on the exposure and resilience of the infrastructure to coastal flooding.

Barriers to Implementation

Some types of infrastructure (e.g., power, water, and communication lines) are relatively easy to move, while larger structures, such as power plants, large highways, and water treatment facilities, are not. These larger structures also require relocating supporting infrastructure, such as feeder pipes, substations, and distribution lines that connect them to the existing infrastructure. In addition, new utility facilities can require large areas of land and the establishment of new utility lines, which sometimes must be routed through private property.

Special Considerations

Regionalization can be an effective way to move critical infrastructure away from flood-prone areas; however, this approach can also increase the overall level of vulnerability, because if the infrastructure is damaged, the impact will be felt over a larger area. Additionally, relocating critical infrastructure components could limit the accessibility for some vulnerable populations.

Fact Sheet A-6: Storm Surge Barriers

A surge barrier is a hard-engineered structure at a river mouth or estuary designed to prevent flooding. It can be either a fixed structure (e.g., closure dam) that is permanently closed or consist of moveable gates or barriers that can be closed when high water levels are forecast.

Cost

Surge barriers tend to have high capital and maintenance costs, although the cost can vary significantly depending on local conditions. In general, movable barriers are more expensive than fixed barriers. Due to the cost, only a handful of movable barriers exist, including the Thames Barrier in London, the Maeslantkering Barrier in Rotterdam, the St. Petersburg Flood Protection Barrier, and the MOSE project in Venice. The costs of these projects varied from $100 million to several billion dollars. Generally, surge barrier construction costs range between $0.7 and $3.5 million per meter, and annual maintenance costs can be about 5 to 10 percent of the capital cost.[6] By preventing the inflow of water to an estuary, closure dams can significantly impact the environment by altering water salinity, temperature, suspended matter, and nutrients, all of which have the potential to affect the local ecosystem. Closure dams also may interfere with the use of the waterways for transportation. Using movable barriers mitigates both these impacts.

Effectiveness

Surge barriers prevent storm surges from entering interior areas for water-level increases up to what they are built to withstand. The technology effectively reduces the height of extreme water levels in the area behind the barrier. For movable barriers, a storm surge monitoring and forecasting system must be used to ensure that the barrier is moved into position before a storm surge arrives.

Barriers to Implementation

Significant environmental costs as well as impacts on waterway transportation can be substantial barriers to implementation.

Special Considerations

Many closure dams are outfitted with turbines to produce electricity from tidal energy. One concern is that closing barriers might cause additional flooding behind the closure during heavy rainfall because the water would not be able to escape the protected area.

---

[6] http://www.unep.org/pdf/TNAhandbook_CoastalErosionFlooding.pdf

Fact Sheet A-7: Beach Nourishment

As beaches erode, coastlines move inland, closer to people and property along the shore, and storm surges are more likely to flood coastal areas. Narrower beaches also expose dunes to wave action and erosion, which negates the protection they offer to coastal infrastructure.[7] Beach nourishment, also known as beach replenishment or renourishment, is a common strategy used to combat such erosion and flooding along sandy coastlines. It involves dumping or pumping high-quality sand from an outside source to replace sand lost to erosion. Nourishment typically uses dredges, trucks, or conveyor belts to move sand from one location to the other. By replacing sand lost to erosion, beach nourishment creates a wider beach, which can help mitigate potential damage to coastal property, It can also cause waves to begin to break farther from the shoreline, weakening their strength before they reach the beach.

Cost

Beach nourishment is a fairly expensive mitigation measure, generally costing between $300 and $1,000 per linear foot, including material, transportation, and construction costs. [8] In addition, the maintenance costs of beach nourishment are substantial, as it must be repeated every few years. Negative environmental impacts are a concern, primarily ecosystem impacts from dredging and burying organisms with the new sand in the area where the nourishment occurs.

Effectiveness

Beach nourishment is a short-term solution that protects people and property by decreasing the energy of waves and limiting how far inland storm surges travel. Beaches must be supplemented with additional quantities of sand every few years, however, for this measure to continue to be effective. Beach nourishment is very effective against water-level increases up to the beach height; for larger events that greatly exceed the beach height, however, beach nourishment will have a minimal effect on mitigating coastal flooding, and the flooding levels will be similar to what they would be without the measure.

Barriers to Implementation

Initial cost and frequent maintenance typically are the largest barriers to implementation. Finding adequate and suitable sand may be a challenge in some areas. Additionally, the question of who pays and who benefits may lead to some outcries about fairness in terms of spending money to protect and potentially increase the value of a small number of beachfront properties. The permitting process could also be slow, given the potential for negative environmental impacts.

Special Considerations

Beach nourishment expands the size of beaches, which could lead to increases in beach-related recreation and associated tourism revenues. Nourishment can have an unintended effect of creating a "false sense of security" for existing or new property owners who underestimate the vulnerability of areas landward of the beach.

---

[7] http://www.iwr.usace.army.mil/docs/projects/HowBeachNourishmentWorksPrimer.pdf
[8] Dean, Robert G. (2002). *Beach nourishment: theory and practice.* Vol. 18. World Scientific Publishing Company.

Fact Sheet A-8: Seawalls

Seawalls are vertical or near-vertical structures built along the coast and designed to prevent erosion and coastal flooding. Seawalls form a protective wall in front of coastal structures and may be constructed from a variety of materials, including concrete, steel, wood, and boulders.

Cost

Seawall construction costs range from $150 to $4,000 per linear foot, depending on engineering and construction specifications and local site conditions.[9] In general, the taller and wider the seawall, the more expensive it is to construct. Seawalls also require significant maintenance over the life of the structure. Seawalls can decrease tourism and accompanying revenue, however, if the associated beach is lost.

Effectiveness

To the degree that a seawall is tall enough and strong enough, it can prevent coastal flooding from sea-level rise and storm-surge. If the water-level increase exceeds the height of the seawall, however, substantial coastal flooding will occur, and the seawall might only reduce wave power and water velocity, not flood height.

With proper construction and maintenance, seawalls last for 30 years, on average, depending on type of construction and the frequency of extreme weather events.[10]

Barriers to Implementation

Seawalls require a high capital cost. Additionally, seawalls often elicit a mixed response from local communities because they can actually increase erosion of fronting beaches and adjacent properties and often generate controversy over their aesthetic impacts. Some areas, by regulation, strictly limit or forbid the use of seawalls.

Special Considerations

Seawalls can cause increased erosion (called "flanking erosion") in adjacent areas. At either end of a seawall, waves tend to reflect sideways along the shore, causing those areas to erode faster.

---

[9] Yohe, Gary, et al. (1996). "The economic cost of greenhouse-induced sea-level rise for developed property in the United States." *Climatic Change* 32(4): 387–410.

[10] Seachange Consulting (2011). *Weighing Your Options: How to Protect Your Property from Shoreline Erosion: A Handbook for Estuarine Property Owners in North Carolina*. North Carolina Division of Coastal Management, National Oceanic and Atmospheric Administration, and Nicholas Institute for Environmental Policy Solutions.

Fact Sheet A-9: Levees and Dikes

Levees, also known as dikes, are constructed embankments designed to reduce the risk of flooding to the areas behind them. Levees are typically built parallel to the course of a river or along coastlines in order to contain, control, or divert the flow of water. Levees are constructed from compacted soil or artificial materials such as concrete or steel. To protect against erosion and scouring, earthen levees can be covered with grass and gravel or a hard surface like stone, asphalt, or concrete.

Cost

When land acquisition costs are low, levees are often the cheapest form of hard defense relative to how much protection they can offer. In general, construction can cost between $100[11] and $1,500 per linear foot depending on the height and slope[12] Maintenance costs are also an ongoing requirement to ensure the structure continues to provide appropriate levels of protection.

Effectiveness

Levees can provide a high degree of protection against flooding in low-lying coastal areas. Effective levees are built with a high volume of material to resist water pressure, sloping sides to reduce wave energy, and crest heights sufficient to prevent overtopping by flood waters. Levees prevent flooding until they are overtopped with water or are breached (i.e., broken or eroded away), at which point flood waters will rise about as high as they would without a levee.

Barriers to Implementation

The high volume and sloping shape of levees necessitates a large building footprint. This factor will be especially important in areas with high property values.

---

[11] FEMA 2007. Selecting Appropriate Mitigation Measures for Floodprone Structures. Table 5.2
[12] Moore. 2009. The Impacts of Sea-Level Rise on the California Coast. Sacramento: California Climate Change Center.

Fact Sheet A-10: Sandbagging

Sandbagging is a simple and common way to prevent or reduce flood damage. A sandbag is a sack made of burlap, polypropylene, or other material that is filled with sand or soil. Sandbags are stacked to build temporary barriers to hold back flood waters. The bags can be brought in empty and filled with local sand or soil.

Cost

Sandbagging is generally one of the cheapest methods for flood protection since the bags cost about 25 cents each and are typically filled by volunteers. The U.S. Army Corps of Engineers estimates that constructing a 100-ft long, 3-ft high sandbag barrier would cost about $3,100.[13] Depending on the situation, additional costs may be incurred for the fill material and removal. Sandbagging can be time-consuming, however, and requires a significant amount of labor.

Effectiveness

Sandbags are usually effective at temporarily mitigating flood damage. They are most effective when used to build barriers that are less than 4 feet tall.[14] Taller or poorly constructed barriers can be prone to sliding and overturning. Sandbags are also susceptible to water seeping through or beneath the bags. Sandbags should be used to provide a basic level of protection, but they are unlikely to hold back flood waters entirely.

Barriers to Implementation

Sandbagging requires a significant amount of labor that must be mobilized quickly before a storm.

Special Considerations

Disposing of sandbags after a flood can be problematic. Sandbags that have been exposed to water cannot easily be reused. Wet sandbags tend to start deteriorating after a few weeks and may be contaminated by unsafe materials in the flood water. Disposing of the bags requires a considerable effort and generates a significant amount of solid waste.

---

[13] http://publications.gc.ca/collections/Collection/D82-46-1999E.pdf
[14] http://www.slate.com/articles/news_and_politics/explainer/2008/06/the_25cent_flood_protection_device.html

Fact Sheet A-11: Elevated Development

Elevated development involves physically raising infrastructure (e.g., on stilts/pilings or raised land) so that water can temporarily flow underneath and/or around without harming the main structure.[15] Elevated development can be included in the original design or added as a retrofit. Traditionally, only buildings are elevated, while the surrounding infrastructure (e.g., roads, walkways) is not; thus, while a building may be protected from flood damage, access to it may be limited during a coastal flood. It is possible to raise surrounding infrastructure, including roads, bridges, walkways, and utility lines. One common example of elevated development is beach homes that are built on stilts, often with the first floor at a height of 10 feet or more above ground level. Elevating structures is a relatively easy feature to incorporate into the design of a facility or infrastructure during initial construction, but it is more challenging to incorporate as a retrofit. Physically raising a structure that is already elevated slightly (e.g. with a crawlspace) is more feasible than elevating "slab-on-grade" construction.

Cost

Incorporating elevated development into the design of a new house adds $2,000 to $30,000 to the cost of the house depending on its size and foundation type, while raising an existing building can easily be well over double this cost. The total costs for raising a building increase with building size and weight but are not directly related to building size.[16, 17] Raising an area of land and/or raising surrounding infrastructure is much more expensive. The cost will vary depending on the amount and type of infrastructure being raised. Raising entire areas of land is very expensive because large amounts of dirt and fill must be transported to the raised site.

Effectiveness

Elevated development is effective in protecting buildings and infrastructure from floods at water levels lower than the base of the first floor of the raised facility and is generally effective for the expected life expectancy of the structure, which might range from 25 to 50 years.

Barriers to Implementation

There are rarely major barriers to implementing elevated development into new construction. Cost to implement, particularly for larger and older buildings, as well as the potential for structural damage, are major barriers to raising an existing building. Creating sufficient access for the handicapped and elderly is also a potential concern. Potential legal liability can hinder being able to raise roads in existing developed areas, because the government entity raising the road is then responsible if drainage patterns are altered and increased flooding results.

---

[15] http://www.usace.army.mil/Missions/CivilWorks/ProjectPlanning/nfpc.aspx
[16] http://www.nytimes.com/2005/12/12/national/nationalspecial/12flood.html
[17] Deschapelles, Natalie (2012). Dissertation: *An Evaluation of, and Suggestions for, Charlotte County Coastal Management, with Regards to Sea Level Rise Vulnerability.*

Special Considerations

Unless the surrounding infrastructure is also raised, flooding will still impact the accessibility of the raised structures and disrupt the economic activities they support.

Fact Sheet A-12: Floating Development

Floating development builds structures on foundations that, when flooded, rise vertically on top of flood waters instead of being inundated. The structures are prevented from moving horizontally by pilings or similar anchors that keep them in the same location and prevents them from floating away. Typically, only individual buildings are constructed on floating foundations, not the surrounding infrastructure. Although there are few common and widely used examples of floating buildings, architects have developed designs that add minimal cost to new construction. Retrofitting an existing facility with a floating foundation would prove challenging, and success would depend on the structure size, age, and type of foundation. Floating development becomes more difficult for buildings with larger weight-to-foundation-size ratios and taller buildings due to the potential instability of a floating foundation.[18]

Cost

Although not a widely used technology, floating development is expected to cost about the same as elevated development. Retrofitting a large and/or old building is often cost prohibitive compared to demolishing and rebuilding, unless the building has historic significance or some other intangible value.[19]

Effectiveness

Floating development is an unproven technology for modern and dense development. Theoretically, it is effective for protecting buildings from floods at water levels lower than the maximum height to which the building is designed to float. Floating development is designed to be effective for the life cycle of the building.

Barriers to Implementation

Theoretically, there are few barriers to incorporating floating development into new construction. The main technological constraints are related to building size. Retrofitting an existing building with a floating foundation can be challenging. The process of setting a building on a new and potentially unstable foundation can result in structural damage.[20]

Special Considerations

Floating development is susceptible to damage from wind, waves, and currents from storms. Therefore, it is only effective in relatively protected areas, not in open water along the coastline, which is susceptible to high-energy waves.

---

[18] http://green.blogs.nytimes.com/2009/10/27/as-sea-levels-rise-dutch-see-floating-cities/
[19] http://www.pbs.org/newshour/rundown/2012/05/preparing-for-a-life-on-water-with-floating-architecture.html
[20] http://www.spur.org/publications/library/report/strategiesformanagingsealevelrise_110109

Fact Sheet A-13: Floodable Development

Buildings can be designed to allow water to flood the lower floor(s) while causing minimal structural damage to the walls, floors, and foundation. On a larger scale, floodable development can also include structures and green infrastructure designed to capture, retain, and gradually release water when the flood recedes. This helps guide flood water away from more sensitive infrastructure. Examples include large underground storage systems such as parking garages designed to hold flood waters, as well as low-lying fields, ponds, or lakes designed to hold water drained from surrounding areas.

Cost

Floodable development can be integrated into facilities at little to no cost for new construction. Costs for neighborhood-level infrastructure to support floodable buildings (for example, a parking garage designed to hold flood water) vary greatly depending on the type and size.

Effectiveness

Floodable buildings are designed to be effective at preventing structural damage from floods up to a specific height. Supporting floodable infrastructure, such as a water-holding parking garage or pond, can hold a specific amount of water, making it easy to calculate how well the structure will mitigate flooding.

Barriers to Implementation

Design-related challenges are associated with floodable buildings because structures must be engineered to withstand regular flooding. Depending on the size of larger scale retention structures, coordination with multiple public and private land owners as well as other stakeholders may be required to work through large-scale planning or land-use issues.

Special Considerations

Floodable development can be hazardous to health, particularly in seaward areas, because floodwater often contains high concentrations of organic chemicals, heavy metals, bacteria, and sediment and could leave behind contamination after the waters subside.[21]

---

[21] http://www.spur.org/publications/library/report/strategiesformanagingsealevelrise_110109

Fact Sheet A-14: Movable Buildings

Movable buildings are designed to be easily relocated in advance of flooding or storm events. The most common movable buildings are trailers and modular buildings, which are moved by truck or train. These buildings are usually left on trailers or set on a concrete slab foundation. Movable buildings are usually limited to small structures because of the logistics and cost involved in relocating them.

Cost

Prefabricated movable buildings, such as modular homes, are inexpensive to build and relocate relative to the cost of replacement. Costs for both construction and relocation increase significantly with size. Large buildings are generally not designed to be movable because the high cost and complicated logistics involved in regularly moving large buildings makes them cost prohibitive—especially when including the costs for multiple moves. In addition, retrofitting buildings to be movable can be expensive compared to demolishing and rebuilding when you add in the cost of relocation.

Effectiveness

The ability to physically move buildings outside of an at-risk area eliminates damage to buildings from flooding. The logistics of moving a large number of buildings are complex, however, because of the time involved. Being able to move buildings requires advance flood warnings as well as sufficient transportation infrastructure to support the movement of both people and buildings.

Barriers to Implementation

There must be designated receiving areas where the buildings will be relocated. These areas must be out of the flood zone but close enough to allow the buildings to be quickly relocated. In addition, planners will want to consider whether utilities and infrastructure are in place to support the buildings for both temporary and permanent relocation.

Special Considerations

Advance planning and early warning are needed to mobilize and coordinate building relocation in advance of a storm or flood.

Fact Sheet A-15: Drainage Systems

Drainage systems can be designed to mitigate flooding on a building, local, or regional level. Like floodable development, buildings and infrastructure can be designed with drains, pumps, and other systems to transport flood water into existing sewer systems, waterways, or other designated floodable areas. Regional drainage systems can include natural or manmade waterways that quickly remove water from a flooded area. Drainage systems are often integrated into stormwater and wastewater infrastructure at the local and regional level; thus, simply diverting large amounts of water into these systems risks overwhelming them, increasing erosion rates, and amplifying the severity of downstream floods.

Cost

Costs for drainage systems vary according to the area over which they are deployed and the size of flood they are designed to mitigate. Large-scale structural drainage systems, typically composed of concrete and steel, have a high cost that increases relative to the amount of area they are supposed to drain and have a cost on par with other heavily engineered structures. Localized drainage systems that simply drain water from a building or parcel into existing waterways are less expensive to install. In addition, manmade drainage systems require regular maintenance to inspect, repair, and clean the systems after large floods.

Effectiveness

While building, parcel-sized, and more localized drainage systems can be designed to accommodate specific-size flood events, large and more regional systems are hard to design, integrate, and maintain. As such, the overall effectiveness of these systems is not very predictable.

Barriers to Implementation

Drainage systems are straightforward to incorporate into development on a local and neighborhood level. Designing effective regional drainage systems can be complex, especially for densely populated areas.

Special Considerations

Trends for stormwater management increasingly focus on managing flooding and stormwater in place through infiltration and evaporation instead of draining into heavily constructed stormwater infrastructure that quickly moves the water off site. In addition, it is important to consider the downstream impacts of draining off large amounts of water and ensure that draining one area does not lead to damaging floods and erosion downstream.

Fact Sheet A-16: Floodproofing Infrastructure

Floodproofing infrastructure uses a combination of methods, including many described in this report, to mitigate or eliminate the potential for flood damage at a given level of flooding, thus lowering the depth-damage curve for a given building. Floodproofing can allow water to flood a building or area (wet flood-proofing) or prevent water from entering a building or area (typically referred to as dry flood-proofing). Building-level examples of floodproofing include sump pumps, spray-on cement, waterproofing membrane, wood or metal shielding, and locating utilities and key equipment at higher levels.

Cost

Floodproofing a building is relatively inexpensive compared to other mitigation measures and usually increases the cost of new construction by 0 to 25 percent, depending on the desired level of protection.

Effectiveness

Usually floodproofing is designed to reduce but not completely eliminate the amount of damage for a given level of flooding. Floodproofing involves the recognition and acceptance of certain amounts of damage for specific flood levels and is only effective up to these predetermined levels. Dry floodproofing is effective up to a certain water-level increase, after which the force of the water acting on the structure can cause it to collapse.

Barriers to Implementation

Cost is typically the largest barrier to implementing floodproofing infrastructure.

Special Considerations

While floodproofing typically protects against certain water-level increases, these measures do not always account for a high velocity of water rushing in from the ocean, which can destroy structures.

Fact Sheet A-17: Preservation of Open Space

Preserving open space involves setting aside land for limited or no development. This can be accomplished in a variety of ways, including by purchasing the land or through a conservation easement. A conservation easement is a voluntary agreement that limits development on a property while maintaining private ownership of the land. In the context of coastal flood mitigation, open space preservation is used to prevent development in vulnerable areas and mitigate the severity of flooding. Preservation of open space is commonly used in conjunction with other green-infrastructure-based flood mitigation measures.[22]

Cost

Costs vary depending on land value and the method of preservation. The cost of preserving land in a locality through fee-simple acquisition—or direct purchase—is equal to the assessed property value. Easements average 60 percent of the cost of fee-simple acquisition; a more accurate cost can be determined by calculating the present worth of the highest value land use that would be given up with the easement.[23] There are also maintenance costs and the potential loss of property taxes versus what could be generated if the land were to be developed.

Effectiveness

For as long as the open space is preserved, designating vulnerable areas as undevelopable is highly effective at preventing flood-related damage in those areas by reducing the amount of infrastructure that could be exposed. Effectiveness is most often evaluated by modeling the impacts of flooding under different open-space and development scenarios.

Barriers to Implementation

The costs to purchase and manage open space present the most significant challenges to implementation because localities often lack the resources to fund effective open-space preservation programs. In addition, property owners can be unwilling to give up development rights or to sell land in prime waterfront areas.

Special Considerations

Preserving open space can also provides valuable social and environmental benefits based on the use of the open space, which may equal or exceed the flood mitigation benefits.

---

[22] American Planning Association (2006). *Planning and Urban Design Standards.*

[23] http://www.rff.org/RFF/Documents/RFF-Rpt-Kousky%20etal%20GreatLakes%20(2).pdf

Fact Sheet A-18: Zoning in Vulnerable Areas

Zoning ordinances restrict allowable land uses for a defined area or district. Zoning may regulate land use; intensity; density; building height, placement, and bulk; and other development features. In the context of flood planning, zoning can prevent or limit development in exposed areas, ensure that new development does not increase the severity of flooding, and require that new and renovated structures incorporate flood-resilient features. Local ordinances must, at a minimum, comply with federal requirements for developing within floodplains, and many zoning ordinances already include measures related to flood-hazard areas.

Cost

Zoning is a low-cost option. Zoning to mitigate flood-related impacts can be incorporated into a municipality's existing zoning code. The local government bears the cost of developing zoning codes, reviewing development applications for consistency with the code, and enforcing the zoning code. These costs vary by the complexity of the zoning code, the volume of development application, and the area and level of enforcement.

Effectiveness

The effectiveness is directly related to the regulations that the zoning implements and the associated enforcement. For example, preventing any development is highly effective, while requiring raised and flood-proofed structures is moderately effective at preventing future damage.

Barriers to Implementation

Processes for changing zoning ordinances vary by state and locality. Often, proposed zoning amendments must go through a lengthy review process that includes public involvement as well as approval by a local board or city council. Zoning that restricts development is sometimes a controversial and unpopular strategy, and may lead to legal challenges.

Special Considerations

Changes to zoning amendments generally must be consistent with a local comprehensive or master plan, which describes the long-term land use goals for an area. Changes to the zoning code that are inconsistent with the plan could be subject to legal challenge.[24]

---

[24] American Planning Association (2006). *Planning and Urban Design Standards*.

Fact Sheet A-19: Development Fees in Vulnerable Areas

Development fees are one-time charges imposed by local governments on new development projects and are designed to cover costs for infrastructure to support the new development. The purpose of a development fee is to have new development pay for the impact to infrastructure instead of having it all paid for by taxpayers. Development fees are used to pay for infrastructure outside the developed area. In the context of flood prevention, development fees could be used to pay for flood-prevention infrastructure or adaptation strategies to prevent flooding in the developed area, or to mitigate the flood-related impacts on other areas that might be caused by the new development.

Cost

Development fees are a low-cost option for local governments. The local governments pay for developing and implementing the ordinances and administering the system. Developers pay the fees to pay for infrastructure associated with new development. Development fees typically only cover construction costs, while maintenance or operating expenses are usually funded through taxes.

Effectiveness

Development fees are an effective way for local governments to pay for new infrastructure to mitigate potential future impacts of coastal flooding brought about by new development. Thus, the effectiveness of the development fees depends on the success of the chosen infrastructure at mitigating flooding damage.

Barriers to Implementation

Not all local governments can enact development-fees because laws and enabling legislation for these fees vary by state.[25]

Special Considerations

If development fees make the overall cost of development in a particular place higher than in neighboring areas, developers may choose to build in another area or decide not to build at all, which can further increase the effectiveness of the measure in terms of minimizing development in vulnerable areas.

---

[25] American Planning Association (2006). *Planning and Urban Design Standards.*

Fact Sheet A-20: Wetlands

Wetlands are ecosystems that may be saturated with water during all or part of the year and are often found at the boundary between land and water. By absorbing flood waters, wetlands protect inland areas from flooding. Coastal wetlands in the United States have been estimated to provide up to $23.2 billion annually in storm protection services.[26]

Communities can take steps to conserve, enhance, restore, or create wetlands in their area. Restoration—returning a degraded or former wetland to its original condition—is typically easier and cheaper to implement than wetland creation, which involves converting either dry land or open water to a wetland.

Cost

Communities can engage in a variety of activities to conserve, create, enhance, or restore wetlands, and specific costs will vary depending on the scope of the project and local conditions. Land acquisition can be a significant upfront cost. Wetland restoration can cost between $3,500 and $80,000 per acre, excluding land costs, and wetland construction typically costs between $35,000 and $150,000 per acre, excluding land costs.[27] Communities usually face additional costs for planning, long-term management, and monitoring, but maintenance costs are typically very low.

Effectiveness

Wetlands reduce the risk of property damage and loss of life from flooding through a number of mechanisms. Wetlands act as natural reservoirs, storing flood waters and then slowly releasing them, delaying and attenuating peak flood flows. Wetlands also dissipate wave, wind, and storm surge energy through resistance provided by the wetland vegetation. Studies have found that a loss of one hectare (about 2.5 acres) of wetland corresponds to an average increase of $33,000 in storm damage from hurricanes.[26] However, wetlands are less likely to be effective in mitigating the effects of very large flood events, particularly regional floods of long duration.

Barriers to Implementation

Construction of new wetlands may not be well-suited for highly developed areas, where there is often no land available or the land can be costly.

Special Considerations

Wetlands provide a number of other valuable services, including improving water quality, providing fish and wildlife habitats, and maintaining water supplies during drought periods.

---

[26] Anderson, Sharolyn J. and Kenneth Mulder (2008). "The value of coastal wetlands for hurricane protection." *Ambio* 37(4): 241.
[27] http://www.bnl.gov/erd/peconic/factsheet/wetlands.pdf

Fact Sheet A-21: Mangroves

Mangroves are medium-sized trees and shrubs found in tropical and subtropical regions that grow in salt water environments. In the United States, mangroves are primarily found in Florida but also exist in other locations around the Gulf of Mexico. Florida's southwest coast supports one of the largest mangrove forests in the world. By acting as a buffer zone against waves and wind, mangrove forests can save lives and reduce storm-surge-related damage.

Cost

The cost of mangroves can vary widely depending on specific conditions. Land acquisition may be necessary to protect existing mangrove forests. If restoration is needed to maintain mangrove forests, the price of labor and the extent of necessary earth work will dramatically affect costs. In general, mangrove restoration has been estimated to cost between $2,150 and $81,000 per acre, excluding land costs.[28]

Effectiveness

Mangroves can reduce storm surge levels by slowing the flow of water and reducing surface waves. Experts estimate that each kilometer of mangrove forest provides a 5- to 50-cm water level reduction. In addition, surface wind waves are expected to be reduced by more than 75 percent over 1 kilometer of mangroves.[29,30] However, some models indicate that mangrove forests may provide less protection against intense, slow-moving hurricanes.

Barriers to Implementation

Mangroves have a limited geographic range due to climate. The temperate climate found in most of the United States is not suitable for mangrove growth.

Special Considerations

Mangroves offer the additional benefits of nutrient retention and wastewater treatment; they support fisheries by serving as a breeding ground and serve as a home to a variety of other wildlife.

---

[28] http://el.erdc.usace.army.mil/elpubs/pdf/vnrs3-2.pdf
[29] http://coastalresilience.org/sites/default/files/resources/storm-surge-reduction-by-mangroves-report.pdf
[30] http://conserveonline.org/workspaces/naturalcoastalprotection/documents/reports/view.html

Fact Sheet A-22: Oyster and Coral Reefs

Reefs—underwater structures made by living creatures—can protect shoreline property from erosion and storm surges. Reefs absorb wave energy, acting as a natural breakwater, and stabilize bottom sediments. Reef restoration—expanding the size and number of reefs—can help communities protect shoreline property and infrastructure.

In the United States, both coral and oyster reefs have been used to protect coastal areas. However, most coral reefs are found in tropical waters and have limited suitability along the coast of the United States. Oyster reefs, on the other hand, are common along much of the eastern seaboard. Reefs are built by adding material such as small bags of oyster shells, loose oyster or clam shells, riprap, or other suitable substances to the water. The material attracts live oyster larvae, which settle and create a live reef.

Restoring coral reefs generally works in the same way; material is provided for corals to settle on. Limestone and concrete are commonly used as these substrate materials are appropriate for coral recruitment. .

Cost

Depending on size, constructing an oyster reef can cost between a few thousand and several million dollars. On average, oyster reef restoration costs about $100 to $150 per linear foot.[31] Oyster reefs have very low maintenance costs. Oyster reefs are extremely durable and may last for 50 years or longer.

Effectiveness

Reefs function as natural breakwaters that reduce the height and energy of waves hitting the shore. According to the Nature Conservancy, in Louisiana, two oyster reef restoration projects, with a total length of 3.6 miles, are expected to reduce wave height by 51 to 90 percent and wave energy by 76 to 99 percent.[32] This reduces both shoreline erosion and storm surge flooding. Reefs are particularly effective against smaller events but may provide limited protection against erosion and coastal flooding during major storms.[33]

Barriers to Implementation

Reef restoration has been known to cause environmental problems. If done irresponsibly, restoration may introduce invasive species to the ecosystem, placing native species at risk. When constructing a reef it is important to take local conditions into account. Seafood safety concerns mean that oyster restoration is not allowed in some locations.

---

[31] Seachange Consulting (2011). *Weighing Your Options: How to Protect Your Property from Shoreline Erosion: A Handbook for Estuarine Property Owners in North Carolina.* North Carolina Division of Coastal Management, National Oceanic and Atmospheric Administration, and Nicholas Institute for Environmental Policy Solutions.
[32] http://www.nature.org/ourinitiatives/regions/northamerica/oyster-restoration-study-kroeger.pdf
[33] http://water.epa.gov/type/oceb/cre/upload/CRE_Synthesis_1-09.pdf

Special Considerations

Reef restoration offers a number of additional benefits beyond coastal protection. For example, reefs can improve water quality and help maintain a healthy coastal ecosystem, promoting commercial and recreational fishing, recreational boating, and tourism.

Fact Sheet A-23: Living Dunes

Coastal dunes are sand deposits located on the land side of a beach. Dunes protect beaches from erosion and protect shoreline property from storm surges. Dunes are extremely fragile. If they are disturbed by human activity, such as trampling, vehicular traffic, and construction, they become unstable and susceptible to erosion. One of the most common and effective methods of stabilizing coastal dunes is to plant vegetation to create living dunes. Some communities also install fencing and/or use discarded Christmas trees to trap sand that ultimately creates new dunes. Building dune walkovers or designating footpaths can also limit damage from human traffic.

Cost

Creating living dunes is a relatively economical option and can be undertaken at the community level using widely available tools. Locally-appropriate vegetation can be obtained from nursery stocks or from nearby intact dune systems. The most commonly used species in dune transplantation are American beach grass, European beach grass, sea oats, and bitter panic grass. American beach grass is sold in 100-stem bundles for approximately $35 to $55 per bundle.[34] The stems are planted two or three to a hole at 1- to 3-foot intervals, resulting in a cost of under $0.50 per square foot of dune.[35] Plantings can be done by hand or with a mechanical planter.[36] New plantings require periodic watering for the first few months and fertilization for up to two years. Decreased access to the beach due to planting vegetation can result in additional costs such as offsetting the loss in access by enhancing or maintaining other beach access points.

Effectiveness

Dune vegetation, which acts to stabilize loose sand, is the most reliable way to protect dunes from wind and waves. Dune plants are especially effective at trapping and holding windborne sand, promoting dune growth over time. Dune vegetation also decreases the wind velocity near the ground, reducing wind erosion at the sand surface. [37] Dune vegetation root networks can also help to stabilize the dune. Ultimately, the vegetation helps preserve dunes and enhances their ability to protect the coast from erosion and coastal flooding. Dunes are effective in preventing flooding up until the water-level increase exceeds the height of the dune, or until sufficient erosion occurs to cause the dune to collapse

Barriers to Implementation

Dunes can sometimes block views of the beach and projects to create new dunes often require landowners to cede some property rights for dune development.

Special Considerations

Special care must be taken in the selection of plant species to avoid introducing invasive species into the ecosystem.

---

[34] http://www.town.orleans.ma.us/pages/orleansma_parks/whgreport.pdf
[35] http://www.whoi.edu/fileserver.do?id=87224&pt=2&p=88900
[36] http://140.194.76.129/publications/eng-manuals/EM_1110-2-1100_vol/PartV/Part-V-Chap_7.pdf
[37] http://conservancy.umn.edu/bitstream/58856/1/2.5.Olafson.pdf

Fact Sheet A-24: Barrier Island Restoration

Barrier islands are naturally occurring, narrow strips of land that run parallel to the coast. Barrier islands protect inland areas during severe weather events by reducing wind and wave energy and by mitigating storm surges. In many areas, however, barrier islands are eroding at extreme rates. In some places, barrier island shorelines are shrinking by up to 100 feet per year.[38]

Barrier island restoration projects are designed to protect and restore barrier island chains. Restoration projects may incorporate a variety of techniques, including using dredged material to increase island height and width, building hard structures to protect the island from erosion, and using sand-trapping fences to build and stabilize sand dunes.

Cost

Barrier island restoration is a large-scale undertaking and can cost tens or hundreds of millions of dollars. For example, restoring East Grand Terre Island in Louisiana, which included restoring 2.8 miles and 620 acres of barrier shoreline and 450 acres of marsh by dredging 3.3 million cubic yards of offshore material, cost $31 million.[39] In addition, barrier island restoration projects require periodic maintenance to ensure resiliency.

Effectiveness

Barrier islands protect coastal communities from damage caused by waves and storm surge. Studies have indicated that the loss of barrier islands can increase wave height by as much as 700 percent during fair-weather forecasts.[40] Recent observations following hurricanes have shown that areas behind restored barrier islands weather storms much better than nearby areas.[41]

Barriers to Implementation

Cost is typically the largest barrier to restoring barrier islands.

Special Considerations

In many cases, barrier islands help shelter wetlands, so restoring barrier islands can indirectly preserve many of the benefits of coastal wetlands.

---

[38] http://pubs.usgs.gov/fs/barrier-islands/
[39] http://www.nola.com/environment/index.ssf/2012/01/louisiana_releases_50-year_blu.html
[40] Stone, Gregory W. and Randolph A. McBride (1998). "Louisiana barrier islands and their importance in wetland protection: forecasting shoreline change and subsequent response of wave climate." *Journal of Coastal Research* 14(3): 900–915.
[41] http://www.habitat.noaa.gov/abouthabitat/barrierislands.html

Appendix B: Approaches and Tools for Monetizing Impacts

Fact Sheet B-1: FEMA Hazus-MH Flood Model

The FEMA Hazus-MH Flood Model uses GIS technology to monetize losses from flooding for a variety of different impacts. Hazus can operate at three levels depending on the needs and expertise of the user. A Level 1 analysis uses default data and models. In a Level 2 analysis, the user can supply custom data, such as depth-damage curves and infrastructure valuation, to perform a more accurate and detailed analysis. In a Level 3 analysis, the user can import the highest level of custom data, including hydrologic data, to produce the most precise and sophisticated, albeit most resource-intensive, analysis. This tool is not specifically tailored to address sea-level rise; however, using scenarios developed in Step 1, this tool can monetize losses for user-provided levels of flooding.

Impacts Monetized

Primary Impacts: Residential building damage, commercial building damage, special facility damage (potable water, wastewater treatment, oil, natural gas, electric power, and communication systems), essential facility damage (hospitals, fire, police, schools, universities), building content loss, vehicle damage, bridge damage, and crop loss.

Secondary Impacts: Relocation costs, business interruption loss through wage loss, rental income loss, and displaced households (measured quantitatively—not monetized), and debris cleanup (from damaged structures only).

Indirect Impacts: Demand and supply of products, change in employment, and change in tax revenues.

Level of Effort and Expertise Needed

This varies from a Level 1 to Level 3 analysis. To perform a Level 1 analysis, someone unfamiliar with the tool could use it after reading the user guide or participating in some basic training. However, it is recommended that someone with more experience using Hazus assist in performing Level 2 and 3 analyses. The Hazus user guide estimates it takes about one to six months to collect input data for a Level 2 analysis and six months to two years to collect the appropriate input data for a Level 3 analysis.

Process to Perform This Approach

Step 1: Download the free model: Set up an account by calling 1-877-336-2627 and access it online by visiting the FEMA Map Service Center (MSC) at msc.fema.gov.

Step 2: Determine the level of analysis you would like to perform and provide someone to train to use the tool or find someone with the appropriate level of expertise.

Step 3: Gather or create the necessary input data for your community to run the tool. This is minimal for Level 1 and just includes a digital elevation model, which is available through NOAA's

Digital Coast or you can create your own. Level 2 and 3 analyses require collecting a longer list of custom inputs that require a much higher level of effort to locate.

Step 4: Run the model. See the user manual and technical manual for FEMA's Hazus-MH Flood Model for more details.

How This Approach Fits Into the Framework of This Guide

This is a very powerful tool, particularly for those performing a holistic approach to monetizing damage to a wide variety of infrastructure for a given water-level increase. This monetization approach makes less sense if you are monetizing damage for a few priority infrastructure elements because it requires that extra resources be used to monetize damages for the rest of the community, which will not change as a result of moving or reinforcing one piece of infrastructure.

This comprehensive tool assesses the depth of water around infrastructure, as is done in Step 1, Task 3, and monetizes impacts of your no-action scenario, as is done in Step 3, Task 2. It can also assess the depth of water around infrastructure in your action scenario if you provide the new water-level increase associated with implementing your action scenario, as is done in Step 2, Task 2. Finally, if you are performing a Level 2 or 3 analysis, you can provide custom data, including new depth-damage curves for reinforced infrastructure, or updated GIS infrastructure maps for a relocated or raised structure to monetize the damages in your action scenario (as is also done in Step 3, Task 2). This approach could be used to perform an impact assessment or risk assessment.

Limitations

- Default data are older (early 2000s) and incomplete for performing Level 1 analyses.
- Processing time can be several days, and custom data collection can take many months when performing Level 2 and 3 analyses.

Key Resources

Case Study:

- *The Role of Land Use in Adaptation to Increased Precipitation and Flooding: A Case Study in Wisconsin's Lower Fox River Basin* provides an example of Level 2 FEMA Hazus-MH Flood Model analysis estimating future losses from flooding in the Lower Fox River Basin of Wisconsin. This Resources For the Future paper also provides some of the pros, cons, and additional insight involved in using this tool to monetize projected flooding losses.

Fact Sheet B-2: COastal Adaptation to Sea level rise Tool (COAST)

The New England Environmental Finance Center (NEEFC) developed the COastal Adaptation to Sea level rise Tool (COAST), which measures physical damage (based on real estate costs) from storm surge and sea-level rise. COAST uses three-dimensional, structure-level outputs to show building-level damage and to break down the losses based on the contribution of storm surge and sea-level rise. For example, Figure B-1 shows how three-dimensional, color-coded outputs are used to present the extent of damage at the building level. This tool incorporates both baseline losses and losses after implementing hypothetical action scenarios.

Figure B-1. COAST Three-Dimensional Building Damage Output

Impacts Monetized

Primary Impacts: Residential building damage, commercial building damage, special facility damage (potable water, wastewater treatment, oil, natural gas, electric power, and communication systems), essential facility damage (hospitals, fire, police, schools, and universities), and building content loss.

Level of Effort and Expertise Needed

Some users will have success with the Web-based tool without prior experience, while others may need expert help; however, you will likely want some GIS expertise to help you obtain the necessary input data.

Process to Perform This Approach

Step 1: Visit the New England Environmental Finance Center website to find the free Web-based tool.

Step 2: Obtain the necessary input data. This typically includes a base land-elevation dataset defining the area, a vector layer defining the assets and asset values, depth-damage function (DDF) for assets (see Key Resources for Fact Sheet B-3 for some sources of depth-damage curves), probability exceedance curves for storm surge, sea-level rise scenarios, and adaptation strategy.

Step 3: Run the tool.

How This Approach Fits Into the Framework of This Guide

This is a very powerful tool, particularly for those performing a holistic approach where you are trying to monetize damage to a wide variety of infrastructure. If you are comfortable using the tool, it can also be quite useful for the priority infrastructure approach. This comprehensive tool helps you assess the depth of water around infrastructure, as is done in Step 1, Task 3; reassess the depth of water around infrastructure based on your action scenario, as is done in Step 2, Task 2; and monetize the impact of flooding in your action and no-action scenarios, as is done in Step 3, Task 2. It is also tailored to incorporate SLR scenarios and different storm types, allows you to input the cost of your action scenario, and provides some summary cost and benefit outputs, which help you compile your net benefits in Step 4. This approach could be used to perform an impact assessment or risk assessment.

Limitations

- Limited to monetizing structural impact damage and building content loss.

Key Resources

Case Study:

- *COAST in Action: 2012 Projects from Maine and New Hampshire* provides some case studies of how the use of COAST is integrated into local decision-making for some Maine and New Hampshire towns. This includes further information about data requirements, outputs, and limitations of the tool.

Fact Sheet B-3: Overlay Infrastructure-Level Economic Data on Flood-Depth Data

This approach involves overlaying economic data about infrastructure values on data about the depth of flooding in Step 1, Task 3. For the holistic approach, you can do this with GIS flood and economic data layers, which is the same basic methodology used in existing comprehensive tools (FEMA Hazus and COAST). For the priority infrastructure approach, you follow the same basic framework; however, you can often do this without GIS data. This approach involves determining the depth of water around your assets, determining how badly the water damages your assets, and determining the value of your assets. This general three-step approach can also be used to estimate the damage to your community's vehicles. The U.S. Army Corps of Engineers (USACE) provides resources for determining the number of vehicles, distribution of vehicles, and value of vehicles in a community.

Impacts Monetized

Primary Impacts: Residential building damage, commercial building damage, special facility damage, essential facility damage, building content loss, and vehicle damage.

Level of Effort and Expertise Needed

The holistic approach requires a very high level of effort and extensive GIS experience. The priority infrastructure is much less time-consuming and can typically be performed without GIS experience because it only requires elevation data of your selected infrastructure.

Process to Perform This Approach

Step 1: Determine the height of the water around your assets.

Option A: holistic approach

Develop a GIS flood layer containing information about exposed infrastructure and the depth of flooding around each infrastructure for each water-level increase selected. See the NOAA *Mapping Coastal Inundation Primer* for more detailed information about creating these mapping layers.

Option B: priority infrastructure approach

Determine the height of water relative to the base of your infrastructure for a given level of flooding. This typically just requires a digital elevation map or elevation data about your infrastructure, as is outlined in Step 1, Task 3.

Step 2: Determine how badly the water damages your assets.

Create or find depth-damage curves—functions that show the percent of damage to your assets—to determine the relationship between flooding height relative to the base of the infrastructure and percent of damage relative to the total value of the asset. The holistic and priority infrastructure approaches involve slightly different processes for this step:

Option A: holistic approach

Consider using generic depth-damage curves developed by USACE to minimize the effort in creating custom curves for your many types of buildings. The "Key Resources" below link to a variety of depth-damage curves for a range of building types, vehicles, and building content.

Option B: priority infrastructure approach

Consider consulting with local building experts or people familiar with the building content to generate your own curves to provide more precise data. For example, you may have a lot of expensive equipment in the basement, or your building may be relatively flood-proof, which can dramatically affect the depth-damage curves.

Step 3: Determine the value of your assets.

Obtain data with information about the value of each building. Check with your town or city assessor's office to see if you can obtain useful data on the assessed value of buildings. If assessed values are not available, check to see if your community can gain access to building or building content insured values that could also be used. You can calculate the damage amount by multiplying the asset value by the percent damage at a given level of flooding that you obtained from the depth-damage curve. The format of the desired data will differ slightly between the holistic and priority infrastructure approach.

Option A: holistic approach

Work with your GIS expert to obtain an economic data layer that can be integrated with your flooding layer.

Option B: priority infrastructure approach

Work closely with people familiar with the infrastructure to obtain detailed information about the infrastructure and content value.

How This Approach Fits Into the Framework of This Guide

This is an alternative approach to using comprehensive tools in determining the monetized costs of buildings and building content (Step 3, Task 2). This general three-step process is applicable to both the **holistic approach** and the **priority infrastructure approach.** It can be used to perform an **impact assessment** or **risk assessment**.

Limitations

- Infrastructure-level valuation data may be difficult to obtain, particularly in a useful format compatible with GIS use. Ensure that you can obtain such data before becoming too invested with this approach.

Key Resources

Depth-Damage Curve Library:

- USACE's *Generic Depth-Damage Relationships for Residential Structures with Basements*
- USACE's *Generic Depth-Damage Relationships for Residential Structures with No Basements*
- USACE's *Analysis of Nonresidential Content Value and Depth-Damage Data*

- USACE's *Depth-Damage Relationships for Structures, Contents, and Vehicles and Content-to-Structure Value Ratios (CSVR) in Support of the Donaldsonville to the Gulf, Louisiana, Feasibility Study*

- USACE's *Generic Depth-Damage Relationships for Vehicles*

Fact Sheet B-4: Use General Economic and Flood Data to Estimate Primary Damage

This approach uses the same basic three-step approach as monetizing infrastructure-level data; however, it provides some lower-resource alternatives if you don't have appropriate infrastructure-level data or are just looking for a quicker means for developing a preliminary estimate of the monetized damage impacts. For example, you may not have data showing the height of water around your infrastructure, which would require you to estimate a depth-damage function, or you may not have infrastructure-level economic valuation data, which might require you to apply a generic economic value such as average building value per square foot or average price per building. Take advantage of your available economic expertise to help you consider and perform alternatives, but remember that more estimates and generic data typically yield less informative and compelling results.

Impacts Measured

Primary Impacts: Residential building damage, commercial building damage, special facility damage, essential facility damage, and building content loss.

Level of Effort and Expertise Needed

The level of effort varies depending on how detailed your available data are; however, the level of effort required is substantially less than the infrastructure-level approach. Some GIS experience may be necessary, and economic expertise is recommended for this approach.

Process to Perform This Approach

Step 1: Determine the height of the water around your assets, or at least identify exposed assets.
Develop a GIS flood map containing information about which infrastructure elements are flooded, and preferably the depth of flooding around each infrastructure for each water-level increase selected. See NOAA's *Mapping Coastal Inundation Primer* for more detailed information about creating these mapping layers. An even lower-resource approach would just be to estimate the number or percent of buildings that are inundated from a quick view of your flood map.

Step 2: Determine how badly the water damages your assets.
Create or find depth-damage curves—functions that show the percent of damage to your assets (in this case building structure or content)—to determine the relationship between water-level height and percent of damage relative to the total value of the asset. Consider using generic depth-damage curves developed by U.S. Army Corps of Engineers (USACE) (see "Key Resources") for many different building types. If your flood map does not contain information regarding the depth of flooding around infrastructure, you may want to try to estimate the average flood depth around buildings and apply the appropriate depth-damage percent to all buildings.

Step 3: Determine the value of your assets.
Obtain generic economic data about the value of your assets (e.g., average price per square foot, average price per building) that makes sense to apply to your flood map. Only use an estimate of the price per square foot if you can assume how many square feet of the building are flooded. It will

increase the accuracy of your estimate if you can obtain more specific values for different building types, such as the average price per commercial, residential, and industrial building.

How This Approach Fits Into the Framework of This Guide

This might be a preliminary approach to approximating monetized damage to infrastructure using the holistic approach. If you are performing the priority infrastructure approach, however, you should probably spend the time finding infrastructure-level data. This approach could be appropriate to monetize the damage of a few water-level increases if you are performing an impact assessment; however, if you are spending the resources to perform a risk assessment, you would probably want to use a more robust approach to monetize primary impacts.

Limitations

- Accuracy: this is more of a "back of the envelope" approach that is only recommended for preliminary assessments or in the absence of more specific data.

Key Resources

Depth-Damage Curve Library:

- USACE's *Generic Depth-Damage Relationships for Residential Structures with Basements*
- USACE's *Generic Depth-Damage Relationships for Residential Structures with No Basements*
- USACE's *Analysis of Nonresidential Content Value and Depth-Damage Data*
- USACE's *Depth-Damage Relationships for Structures, Contents, and Vehicles and Content-to-Structure Value Ratios (CSVR) in Support of the Donaldsonville to the Gulf, Louisiana, Feasibility Study*

Fact Sheet B-5: Monetize Business Interruption Loss

Business interruption loss is typically caused by inaccessible roads, lack of essential utilities, or damage to buildings or equipment. The basic approach for determining the monetized loss is to estimate how long a business is inoperable and multiply that by an economic indicator of the business's output such as wages or sales. While this might be an oversimplification, it provides a general approach to approximate business loss without becoming overly resource intensive. Work with your economist to determine how detailed and robust you want to make this analysis. It is worth noting that the FEMA Hazus-MH Flood Model monetizes business interruption loss through wage loss if you choose to use that comprehensive tool.

Impacts Measured

Secondary Impacts: Business interruption loss.

Level of Effort and Expertise Needed

The level of effort increases as you try to assess infrastructure (business)-level losses. You will need economic expertise for this approach.

Process to Perform This Approach

Step 1: Determine the amount of time a business is inoperable.

For floods where substantial building damage is expected, the loss of business due to damage or destroyed buildings and equipment may be a substantial component of business interruption. See Table 14.12 of the FEMA Hazus-MH Flood Model technical manual for the expected restoration time of various building types due to flooding. This can help you estimate how long a business may be out of operation after a flood event.

It is also possible that the primary business interruption may be caused by inaccessible roads or lack of electricity or water while crews are working to get utilities back up. In this case, you might expect businesses to be inoperable for several days to more than a week while the floodwaters retreat, utilities get back up, and businesses clean up from the flood. In this case, you may want to generate some more general assumptions, perhaps using data from historic storms, for the amount of time you can expect your businesses to be inoperable.

Step 2: Determine the wages or output of a business.

In Table 14.14 of the Hazus-MH Flood Model technical manual, FEMA provides data about the expected wages, income, and sales output per building type per square foot. You can multiply any of these variables by the number of days a business is inoperable to estimate business loss. Additionally, the Bureau of Labor Statistics (BLS) Quarterly Census of Employment and Wages and the Census Bureau County Business Patterns provide data on the number of establishments and employees and total wages by industry type. The BLS data are released sooner and offered at the county level each quarter of the year, while the Census data are annual but offer the advantage of being released at the ZIP Code and county level. You could multiply the wage data by the length of time a certain number or percent of businesses in each industry is inoperable to estimate lost wages due to business interruption.

How This Approach Fits Into the Framework of This Guide

This helps you monetize one of the major secondary impacts (business operation loss) in Step 3, Task 2 of this framework.

Limitations

- The economy is very complex, and in the longer term, businesses may start to shift production to neighboring facilities or workers may find new work, which can often offset some of the business loss impacts.

Fact Sheet B-6: Non-Market Valuation Methodologies

These methodologies measure the value of goods, services, or states of nature not traded on the market. Many of these non-market valuation studies require implementing surveys or gathering substantial data, which is one reason why many environmental impacts, other costs, and benefits are often only considered qualitatively. The methodologies are described below; however, the processes for performing the methodologies are not outlined. Be sure to consult your experienced economist if you choose to perform any of these methodologies.

Description

Travel Cost Method: Uses travel costs such as hotel, transportation, and entrance fees to approximate the value of a coastal resource. This also typically includes a survey of a sample of people who have used the resource to estimate the difference between what the consumer actually paid for the resource and what they would be willing to pay.

Hedonic Price Method: Models how the price of a private good, such as real estate, relates to the attributes that might affect price, such as beach proximity and flooding resilience. To use this method in the context of coastal flooding, you would look at how one attribute (e.g., an elevated house better protected from coastal flooding) would change the price of the house.

Willingness to Pay: Surveys how much people say they would be willing to pay for an environmental improvement or to avoid a decrease in environmental quality.

Avoided Cost Methodology: Estimates the economic value of an environmental resource or service based on additional costs that would be placed on society if the resource or service no longer existed.

Impacts Measured

These methodologies can be used to measure just about any **environmental cost, other cost,** or **benefit.**

Level of Effort and Expertise Needed

These methodologies typically involve somewhat resource-intensive surveys or data collection and would require someone with a high level of economic expertise.

How These Approaches Fit Into the Framework of This Guide

These methodologies help you monetize environmental impacts, other costs, and benefits in Step 3, Task 2 of this framework.

Key Resources

- NOAA's *Introduction to Economics for Coastal Managers* further explains many of these approaches and provides several case studies.
- The National Ocean Economics Program's Environmental & Recreational (Non-Market) Values— Research Methodologies provides additional details about these and many other non-market valuation techniques.

Fact Sheet B-7: Benefits Transfer

Benefits transfer is an approach that borrows economic values from one study to apply to your own parallel situation. It is a reasonably low-resource method for monetizing impacts because you are not implementing the methodology for collecting the local data; however, it typically provides less informative and compelling data than gathering data specific to your local situation.

Impacts Measured

This approach can be used to measure just about any type of **primary impact, secondary impact, environmental impact, other cost,** or **benefit**. It is best used to monetize smaller impacts that you might otherwise not have the resources to monetize at all; therefore, you might want to avoid using this method to monetize your most substantial impacts.

Level of Effort and Expertise Needed

Benefits transfer typically requires a low level of effort. It should include someone with some basic economic understanding.

Process to Perform This Approach

Step 1: Find existing studies monetizing the specific type of impact you want to monetize in your community.

Step 2: Evaluate how well the studies transfer to your situation. Try to find one with many similarities to your local conditions.

Step 3: Evaluate the quality of the study from a research and methodology perspective.

Step 4: Use the previous two steps to identify the best study. Consider tailoring the value to your community by adjusting the monetized value based on differences in the area of study and your community.

How This Approach Fits Into the Framework of This Guide

This helps you monetize just about any impact for which you find it too resource intensive to create your own study in Step 3, Task 2 of this framework.

Limitations

The more similar the study's local conditions are to your own community, the more reliable the estimate will be. However, it is less informative and compelling than gathering local data.

Key Resources

- *Introduction to Economics for Coastal Managers* includes some case studies using benefits transfer.
- The National Ocean Economics Program's benefit transfer website provides additional details about this methodology as well as several case studies.

Fact Sheet B-8: Input-Output Models to Measure Indirect and Induced Impacts

Indirect impacts are a measure of the value of the inputs used by firms that are called on to produce additional goods and services for those firms first impacted by coastal flooding. For example, if hotels are damaged and inoperable for a period of time due to coastal flooding, firms that normally provide food and other supplies to the hotel will not be doing so, while some construction firms may increase their output by fixing the building. The indirect impacts are these increases or decreases in output of the firms that service the entities affected by coastal flooding.

Induced effects are related to individuals and businesses that receive added or reduced income as a result of local spending by firms and plants that are affected by the direct and indirect impacts of coastal flooding. For example, coastal flooding may result in reduced tourism for a period of time and negatively affect local hotels. Employees of these hotels might receive decreased pay as a result, and the induced impacts are the reduced sales, jobs, and outputs in the local economy as a result of less tourist money flowing in.

Typically, indirect and induced impacts are relatively small compared to the primary impacts of coastal flooding; thus, based on your resource availability, you may choose to consider this type of modeling as a supplementary type of analysis if the results are still very unclear after monetizing the primary impacts, secondary impacts, environmental impacts, other costs, and benefits. The primary tools that can be used to perform this input-output indirect and induced modeling include the Impact Analysis for Planning (IMPLAN) and the Bureau of Economic Analysis Regional Multipliers from the Regional Input-Output Modeling System (RIMS II).

Impacts Measured

Indirect and Induced Impacts

Level of Effort and Expertise Needed

These models require a high level of effort, need to be purchased, and require someone with economic modeling experience.

Process to Perform This Approach

Step 1: Identify someone with expertise in using these input-output models.

Step 2: Purchase and download the models. The links are shown in the "Key Resources" section below.

Step 3: Run the models. Consult user guides and your available economic expertise accordingly.

How This Approach Fits Into the Framework of This Guide

These input-output models help you monetize indirect and induced impacts in addition to the types impacts monetized Step 3, Task 2 of this guide.

Limitations

These models are technical and resource-intensive. Depending on the results of your analysis, the additional monetized impacts from using these models may or may not impact the ultimate decision. Thus, consider using these tools for supplementary analysis when resources allow.

Key Resources

- IMPLAN: Further information and how to order.
- BEA's RIMS II: Further information, how to order, and access to the user guide.

Appendix C: Relevant Case Studies

Title and Organization	Geographic Area	Methodology and Key Findings	Keywords
		United States Case Studies	
Climate Change and Coastal Flooding in Metro Boston: Impacts and Adaptation Strategies [42] Journal: Climate Change	New England	*Used GIS to model flood elevation versus flooded area curve, and used bootstrapping of historical sea-level data and Monte Carlo simulation to project sea levels and impacts for the next century. Used per linear meter cost from U.S. Army Corps of Engineers to estimate cost of resilient infrastructure.* • Performed risk-based analysis to show the cumulative 100-year economic impacts on developed areas from increased storm surge due to sea-level rise. • Concluded that it is advantageous to use expensive structural protection in highly developed areas and land management approaches in less developed or environmentally sensitive areas. • Determined that approximately 40 percent of the residences in the 100-year floodplain and 25 percent in the 100- and 500-year floodplains get flooded during an event in the Boston Metro area. • Determined that damages averaged between $7,000 and $18,000, depending on location.	Grey Infrastructure, Seawalls, Metropolitan Areas, Inundation, Property Damage, GIS, Land Use Management, Climate Change, Sea-Level Rise

[42] Kirshen, Paul, K. Knee, and M. Ruth. (2008). Climate Change and Coastal Flooding in Metro Boston: Impacts and Adaptation Strategies. *Climatic Change* 90:453–473.

What Will Adaptation Cost?
An Economic Framework for Community Planners

June 2013

Appendix C: Relevant Case Studies

Title and Organization	Geographic Area	Methodology and Key Findings	Keywords
COAST in Action: 2012 Projects from New Hampshire and Maine[43] New England Environmental Finance Center	New England	Used the COastal Adaptation to Sea-Level Rise Tool (COAST) • Estimates the economic costs and benefits of adaptation—specifically regarding the use of sea walls—under various sea-level rise and adaptation scenarios in several New England towns. • Assessed impacts of three sea-level rise scenarios (none, low, and high). • Includes detailed estimates from architectural engineering firms broken down by labor and equipment. • Assumed a discount rate of 3.5 percent, or 1 percent higher than projected inflation.	Sea-Level Rise, Storm Surge, Inundation, Grey Infrastructure, Flood Walls

[43] Merrill, S., P. Kirshen, D. Yakovleff, S. Lloyd, C. Keeley, and B. Hill. (2012). COAST in Action: 2012 Projects from New Hampshire and Maine. New England Environmental Finance Center Series Report #12-05. Portland, Maine.

Appendix C: Relevant Case Studies

Title and Organization	Geographic Area	Methodology and Key Findings	Keywords
Simplified Method for Scenario-Based Risk Assessment Adaptation Planning in the Coastal Zone[44] Journal: Climate Change	New England	*Used scenario-based risk assessment based on developing exceedance curves of flood elevations. This methodology was incorporated into the more user-friendly Coastal Adaptation to Sea level rise Tool (COAST).* • Investigated beach nourishment as a mitigation option for sea-level rise and beach erosion. • Concluded that beach nourishment up to the present 100-year flood elevation is estimated to cost $3,280 per linear meter ($1,000 per linear foot). • Assumes that 75 percent of the nourished beach must be redone every 10 years. • Calculated estimates of flood damages to buildings and contents using the generic depth-damage relationships for residential structures with basements from the U.S. Army Corps of Engineers, which are based on building replacement values derived from assessors' tables. • COAST is intended for local and regional planners, municipal officials, university extension agents, and other decision-makers to help them understand the potential economic impacts of different scenarios of sea-level rise and the associated costs of different adaptation strategies.	Beach Erosion, Beach Nourishment, COAST tool

---

[44] Kirshen, Paul, et al. (2012). Simplified method for scenario-based risk assessment adaptation planning in the coastal zone. Climatic Change 1–13.

Appendix C: Relevant Case Studies

Title and Organization	Geographic Area	Methodology and Key Findings	Keywords
Assessing Future Risk: Quantifying the Effects of Sea Level Rise on Storm Surge Risk for the Southern Shores of Long Island, New York[45] Journal: Natural Hazards	Mid-Atlantic	*Used hazard analysis tools such as FEMA's HAZUS tool, USGS's asymmetric mapping tool, storm surge impact zone developed via National Hurricane Center's SLOSH model and builds upon NOAA's Community Vulnerability Assessment Tool (CVAT) methodology.* • Followed a risk assessment approach that integrated and expanded on the functionality of readily available, low-cost, or free, frequently used tools. • Outlined steps for measuring and mapping exposure, vulnerability, and overall risk from a category 3 storm surge in Suffolk County on Long Island, New York. • Examined elevated risk when factoring sea-level rise as projected by 2080.	Storm Surge, Sea-Level Rise, Exposure Index, GIS, Publicly Accessible Tools
Risk Increase to Infrastructure Due to Sea Level Rise. Climate Change and a Global City: An Assessment of the Metropolitan East Coast Region[46] Columbia Earth Institute for the U. S. Global Change Research Program	Mid-Atlantic	*Used the FEMA flood insurance map to identify vulnerable areas of the city. Used total costs from past category 1–4 storms, and scaled benefit of mitigation options based on asset density in the New York City metropolitan area.* • Provided a general risk assessment of the potential damage of category 1–4 hurricanes that could hit New York over the next 100 years. • Looked at overall insured value of New York City, used total losses from past storms in other areas, and projected sea-level rise to estimate what losses would be in New York for a comparable storm. • Evaluated types of grey infrastructure, such as sea walls and dykes, and also weighed land-use management and retreat as viable mitigation options.	Sea-Level Rise, Storm Surge, Hurricanes, Metropolitan Area

[45] Shepard, Christine C., et al. (2011). Assessing Future Risk: Quantifying the Effects of Sea Level Rise on Storm Surge Risk for the Southern Shores of Long Island, New York. *Natural Hazards* 1–19.
[46] Jacob, K. H., N. Edelblum, and J. Arnold. (2000). Risk Increase to Infrastructure Due to Sea Level Rise. Climate Change and a Global City: An Assessment of the Metropolitan East Coast (MEC) Region.

Appendix C: Relevant Case Studies

Title and Organization	Geographic Area	Methodology and Key Findings	Keywords
Measuring the Impacts of Climate Change on North Carolina Coastal Resources[47] National Commission on Energy Policy	Eastern Seaboard	*Used high-resolution topographic LIDAR (Light Detection and Ranging) data, MAGICC/SCENGEN Global Climate Model, the hurricane wind speed model (HURRECON), property values data from North Carolina county tax offices, and other available GIS data.* *Used hedonic property model, nested logic demand models, NLOGIT econometric software, National Marine Fisheries Service (NMFS), and Marine Recreational Fishery Statistics Survey (MRFSS) willingness to pay surveys, and extrapolated from data collected from past storm events.* *Used U.S. Army Corps of Engineers "rule-of-thumb" of 1 cubic yard per running foot of beach to estimate cost to replace each foot of eroding beach, at an average cost of $6 per cubic yard for East Coast barrier beaches in 2004 dollars.* - Assessed inundation and storm surge vulnerability for New Hanover, Dare, Carteret, and Bertie Counties of North Carolina. - Estimated potential decrease in residential property values, lost recreation and tourism income, and costs of business interruption, including the agriculture, forestry, and commercial fisheries industries. - Estimated a potential loss of nearly 50 percent in recreational revenues in 2080 due to beach erosion. - Concluded that the economic impacts of sea-level rise on these four coastal North Carolina counties could be significant if not mitigated. - Concluded that annual beach nourishment costs to mitigate sea-level rise from 2004 to 2080 is $348 million per year when discounted at a 2-percent rate.	Green Infrastructure, Inundation, Storm Surge, Sea-Level Rise, Beach Erosion, Hedonic Property Model, Nested Logic Demand Model, Willingness-to-Pay Surveys, Beach Nourishment

[47] Bin, Okmyung, Chris Dumas, Ben Poulter, and John Whitehead. (2007). Measuring the Impacts of Climate Change on North Carolina Coastal Resources: Final Report. National Commission on Energy Policy.

What Will Adaptation Cost?
An Economic Framework for Community Planners

June 2013

Appendix C: Relevant Case Studies

Title and Organization	Geographic Area	Methodology and Key Findings	Keywords
Estimating the Vulnerability of U.S. Coastal Areas to Hurricane Damage[48] Francis Marion University Quality Enhancement Plan	Southeast United States	*Used multi-attribute utility theory(MAUT) to formulate a hurricane vulnerability index (HVI) using seven indicators: population, number of housing units, house value, probability of hurricane strike, building code effectiveness, building age, and vulnerability to sea-level rise. Used 2000 U.S. Census data for population and housing values.* • Calculated an HVI to rank the susceptibility of coastal areas to hurricane damage and provided necessary equations. • Ranked 11 U.S. counties in terms of vulnerability, with Miami-Dade ranking the most vulnerable. • Identified the top 10 U.S. counties in terms of cumulative exposure in coastal Florida (6), North Carolina (3), and Louisiana (1). • Identified three key elements—the level of exposure, the physical susceptibility to the hurricane, and the frequency and intensity of hurricanes—that contribute to the degree to which communities experience hurricane damage. • Considered wetlands, levees, sea walls, and storm-resilient construction as viable infrastructure options. • Stated that coastal Louisiana wetlands provided $452 acre/year in storm reduction benefits.	Vulnerability Index, Sea-Level Rise, Climate Change, Storm Surge, Hurricanes, Event Probability, Grey Infrastructure, Green Infrastructure
An Adaptive Regional Input-Output Model and its Application to the Assessment of the Economic Cost of Katrina[49] Journal: Risk Analysis	Gulf of Mexico	*Uses the Adaptive Regional Input-Output (ARIO) Model to estimate indirect costs. Also provides a methodology for converting national-level BEA input-output tables to regional-level input-output tables for use in the ARIO model.* • Derived regional-level input-output tables from Bureau of Economic Analysis national-level input-output tables to simulate the response of the economy of Louisiana to the landfall of Katrina. • Considers damage from Katrina as a whole—primarily from, but not specific to, inundation.	Hurricane Katrina, Input-Output Models, Storm Surge

[48] Pompe, Jeffrey and Jennifer Haluska. (2011). Estimating the Vulnerability of U.S. Coastal Areas to Hurricane Damage.
[49] Hallegatte, S. (2008). An adaptive regional input-output model and its application to the assessment of the economic cost of Katrina, Risk Analysis.

Appendix C: Relevant Case Studies

Title and Organization	Geographic Area	Methodology and Key Findings	Keywords
The Socio-Economic Impact of Sea Level Rise in the Galveston Bay Region[50] Environmental Defense Fund	Gulf of Mexico	*Used FEMA HAZUS Multi-Hazard MR3 ArcGIS Extension.* • Illustrated the impact that climate change will have on communities in the Galveston Bay region using two scenarios of sea-level rise: 0.69 meters and 1.5 meters. • Estimated impact on displaced populations; residential structures; industrial facilities, hazardous waste, Superfund, and solid waste sites; water treatment plants; and building-related economic losses. • Found that, under the 1.5 meter scenario, almost 99,000 households would be displaced and more than 75,000 structures impacted in the region. • Did not estimate the impact on municipal and county governments, as they deal with infrastructure loss, moving populations, and potentially decreasing tax base. • Looked only at the losses assuming no mitigation is performed.	Gulf of Mexico, Inundation, Storm Surge, Hurricanes, Sea-Level Rise
The Role of Land Use in Adaptation to Increased Precipitation and Flooding: A Case Study in Wisconsin's Lower Fox River Basin[51] Resources for the Future	Great Lakes	*Used HAZUS and national databases of the inventory of structures at the census block level from the U.S. Bureau of the Census and Dun & Bradstreet to estimate depth damage curves, debris generation, indirect losses (such as loss of income or relocation expenses), and displacement of people.* • Examined flooding losses and mitigation benefits for the Lower Fox River basin from central Wisconsin to Green Bay. • Investigated land-use options as an alternative to traditional grey infrastructure mitigation. • Considered traditional grey infrastructure such as reservoirs, levees, flood walls, and dams, as well as land-use regulations for increasing wetland areas and removing or mitigating the influence of drainage ditches in agricultural lands. • Estimated losses, but not benefits, of mitigation options. • Focused primarily on regulating land use as a mitigation option.	Land Use, Inundation, Water Quality, Great Lakes, Wetlands, Recreation, Climate Change, Grey Infrastructure, Green Infrastructure

[50] Yoskowitz, David, et al. (2009). The Socio-Economic Impact of Sea Level Rise in the Galveston Bay Region. Environmental Defense Fund.
[51] Kousky, Carolyn, et al. (2011). The Role of Land Use in Adaptation to Increased Precipitation and Flooding: A Case Study in Wisconsin's Lower Fox River Basin. Resources for the Future.

What Will Adaptation Cost?
An Economic Framework for Community Planners

June 2013

Appendix C: Relevant Case Studies

Title and Organization	Geographic Area	Methodology and Key Findings	Keywords
Potential Impacts of Increased Coastal Flooding in California Due to Sea-level Rise[52] Journal: Climatic Change	California	*Used GIS database with inundation maps combined with FEMA HAZUS; data from the U.S. Census, TeleAtlas, the California Office of Health Planning, and power plants from the California Energy Commission; and information on hazardous materials sites from U.S. EPA.* • Assessed the potential impacts from projected sea-level rise if no actions are taken to protect the coast, focusing on impacts to the state's population and infrastructure. • Discussed beach erosion impacts, effects on wetlands, costs of coastal defenses, and issues of social and environmental justice related to sea-level rise. • Considered the replacement value of structures by intersecting the inundation layers with year 2000 census block data, assuming that building value data is distributed uniformly over a census block's area. • Provided detail on the timing and degree of vulnerability for different levels of rise. • Quoted other existing research and literature for some basic estimates of physical adaptation options, including building or strengthening seawalls and levees.	Sea-Level Rise, Beach Erosion, Environmental Justice, Grey Infrastructure, GIS

[52] Heberger, Matthew, et al. (2011). Potential impacts of increased coastal flooding in California due to sea-level rise. *Climatic Change* 1–21.

Appendix C: Relevant Case Studies

Title and Organization	Geographic Area	Methodology and Key Findings	Keywords
		European Case Studies	
Assessing Climate Change Impacts, Sea Level Rise and Storm Surge Risk in Port Cities: A Case Study on Copenhagen[53] OECD Environment Working Papers, OECD Publishing	Europe/Scandinavia	*Used statistical data available from past storm scenarios, combined with GIS analysis of the population and asset exposure in the city, for various sea levels and storm surge characteristics.* • Estimated economic impacts of climate change on Copenhagen, Denmark, focusing on sea-level rise and storm surge. • Assessed exposure by mapping population and industry distribution data onto a Digital Terrain Model (DTM) (exposure data came from risk management software). • Estimated direct and indirect losses as part of a risk analysis due to climate change and sea-level rise.	Inundation, Storm Surge, Sea-Level Rise, Climate Change, Sea Walls, Risk Analysis

[53] Hallegatte, S. et al. (2009). Assessing Climate Change Impacts, Sea Level Rise and Storm Surge Risk in Port Cities: A Case Study on Copenhagen. OECD Environment Working Papers, No. 3, OECD Publishing.

Appendix C: Relevant Case Studies

Title and Organization	Geographic Area	Methodology and Key Findings	Keywords
Impacts of Sea Level Rise Towards 2100 on Buildings in Norway[54] Journal: Building Research and Information	Europe/Scandinavia	*Used the Norwegian Building Matrix and a custom GIS program to perform vulnerability assessment and develop a model (Rahmstorf [2007]) for future global sea level rise and the "business-as-usual" emissions scenario from the IPCC.* • Assessed the impact of sea-level rise on grey infrastructure along the Norwegian coastline. • Estimated that more than 100,000 buildings are located less than 1 meter above typical sea level, including more than 2,000 hotels and restaurants, 6,000 residences, 500 buildings related to communications infrastructure, 270 buildings vital to energy supply, 160 schools, and 20 health care facilities. • Calculated the total potential costs of the expected sea-level rise by estimating an average cost per building for several building types and then multiplying that estimate by the number of structures per category. • Concluded that improved building codes that encourage structures resistant to natural stressors, increased drainage, installation of pumps, and relocating vulnerable structures could greatly reduce costs.	Sea-Level Rise, Storm Surge, Inundation, Grey Infrastructure, Climate Change, GIS

[54] Almas, Anders-Johan and Hans Olav Hygen (2012). Impacts of Sea Level Rise Towards 2100 on Buildings in Norway. *Building Research and Information* 40:3, 245–259.

Appendix C: Relevant Case Studies

Title and Organization	Geographic Area	Methodology and Key Findings	Keywords
		Asian Case Studies	
An Assessment of the Socio-Economic Impacts of Floods in Large Coastal Areas[55] Asia-Pacific Network for Global Change Research	Southeast Asia	*Used the Urban Flood Risk Analysis (URFA) model, the Institute of Industrial Science Distributed Hydrological Model (IISDHM), and a hydrodynamic model developed by the Public Work Research Institute.* • Examined the socio-economic impact of floods exacerbated by climate change in six countries in South and Southeast Asia. • Developed GIS database of hydrologic characteristics and socioeconomic conditions. • Adapted existing tools and methodologies to simulate and assess flooding under long-term climatic change and rise in sea-level, and estimated socio-economic impacts and vulnerability. • Identified existing policy gaps. • Developed strategies to better communicate with the public. • Did not consider the economic impacts of sea-level rise on ecosystems, tourism, fishing, and other coastal industries.	Land-Use Management, Asia, Climate Change, Sea-Level Rise. GIS, Inundation, Buildings, Industrial and Commercial Vulnerability

[55] Hallstrom, D.G. and V.K. Smith, V.K. (2005). Market responses to hurricanes. *Journal of Environmental Economics and Management* 50: 541–561.

Appendix C: Relevant Case Studies

Title and Organization	Geographic Area	Methodology and Key Findings	Keywords
The Economic Impact of Sea-Level Rise on Nonmarket Lands in Singapore[56] Ambio: A Journal of the Human Environment	Southeast Asia	*Used a travel cost and contingent valuation study to estimate the value of nonmarket land.* • Discussed nonmarket lands, such as beaches, marshes, and mangrove estuaries, and revealed that consumers in Singapore attach considerable value to beaches. • Looked at the cost of protection given different sea-level rise scenarios and the nonmarket value of beaches, marshes, and mangrove estuaries. • The contingent valuation study also placed high value on marshes and mangroves, but this was not supported by the travel cost study. States that, although protecting nonmarket land uses from sea-level rise is expensive, highly valued resources, such as Singapore's popular beaches, should be protected. • Presented method for estimating costs of beach protection, including the construction costs of the underwater hard structures, the cost of sand, and the maintenance costs. The cost of sand is estimated to be $30 per cubic meter, and the maintenance costs are assumed to be 4 percent of construction costs.	Beach Erosion, Beach Nourishment, Nonmarket Valuation, Land-Use Management, Sea-Level Rise, Recreation

[56] Ng, W. and R. Mendelsohn (2006). The Economic Impact of Sea-Level Rise on Nonmarket Lands in Singapore. *Ambio*, 35 (6), 289–296.

Appendix C: Relevant Case Studies

Title and Organization	Geographic Area	*Methodology and Key Findings*	Keywords
Flood Risks, Climate Change Impacts and Adaptation Benefits in Mumbai: An Initial Assessment of Socio-Economic Consequences of Present[57] Organization for Economic Cooperation and Development (OECD)	India/Asia	*Used the Storm Water Management Model (SWMM) to assess vulnerability. Used three types of data to estimate direct costs: 1) actual data from the 2005 flood, 2) insured value data, and 3) average data for generic industrial, commercial, and residential facilities. Used the Adaptive Regional Input-Output (ARIO) model to estimate indirect costs.* • Provided summary information for unprecedented flooding experienced in Mumbai, India, in 2005. The event caused direct economic damages estimated at nearly $2 billion (U.S. dollars) and caused 500 fatalities. • Discussed relationship between direct and indirect costs. • Concluded that by the 2080, in a SRES A2 scenario (an "upper bound" climate scenario), the likelihood of a 2005-like event would more than double. • Estimated that total losses (direct plus indirect) associated with a 1-in-100 year event could triple due to climate change. • Investigated the benefits of upgrades to the city's drainage system. • Did not take into account population and economic growth. • SWMM model is not applicable to typical sea-level rise and storm-surge vulnerability assessments.	Direct Costs, Indirect Costs, Grey Infrastructure

[57] Hallegatte, S. et al. (2010), "Flood Risks, Climate Change Impacts and Adaptation Benefits in Mumbai: An Initial Assessment of Socio-Economic Consequences of Present and Climate Change Induced Flood Risks and of Possible Adaptation Options," OECD Environment Working Papers, No. 27, OECD Publishing.

Appendix C: Relevant Case Studies

Title and Organization	Geographic Area	Methodology and Key Findings	Keywords
Predicted Impact of the Sea-level Rise at Vellar-Coleroon Estuarine Region of Tamil Nadu Coast in India: Mainstreaming Adaptation as a Coastal Zone Management Option[58] Journal: Ocean & Coastal Management	India/Asia	*Used the Digital Elevation Model (DEM) derived from SRTM 90M (Shuttle Radar Topographic Mission) data, along with GIS techniques, to identify an area of inundation in the study site. Effects of inundation are discussed using GIS-based land-use land-coverage (LULC) and hamlet mapping.* • Assessed the impacts of sea-level rise on coastal natural resources and dependent social communities in the low-lying area of Vellar-Coleroon estuarine region of the Tamil Nadu coast, India. • Discussed a number of types of grey and gray infrastructure (but these are not incorporated in the model). • Calculated vulnerability of coastal areas to inundation based on projected sea-level rise scenarios of 0.5 and 1.0 meters.	Sea-Level Rise, Grey Infrastructure, Green Infrastructure, GIS
Other Regional Case Studies			
Assessing the costs of sea-level rise and extreme flooding at the local level: A GIS-based approach[59] (26) Journal: Ocean & Coastal Management	Israel	*Used GIS modeling to create inundation increment maps and spatially distributing assets and population to generate a comprehensive picture of local socio-economic costs.* • Used a method of performing vulnerability assessment that is generic and transferable. • Estimated costs related to different scenarios of permanent inundation and periodic flooding. • Inventoried various tools available for assessing the impacts of natural processes at the local level. • Identified the various types of capital stocks at risk.	Inundation, Sea-Level Rise, Climate Change, GIS

[58] Saleem Khan, A.; et al. (2012). Predicted Impact of the Sea-level Rise at Vellar-Coleroon Estuarine Region of Tamil Nadu Coast in India: Mainstreaming Adaptation as a Coastal Zone Management Option. *Ocean & Coastal Management*.
[59] Lichter, Michal and Daniel Felsenstein (2012). Assessing the costs of sea-level rise and extreme flooding at the local level: A GIS-based approach. *Ocean & Coastal Management* 59 (2012): 47–62.

What Will Adaptation Cost?
An Economic Framework for Community Planners

June 2013

Appendix C: Relevant Case Studies

Title and Organization	Geographic Area	Methodology and Key Findings	Keywords
The Vulnerability of Caribbean Coastal Tourism to Scenarios of Climate Change Related Sea Level Rise[60] *(41)* Journal of Sustainable Tourism	Caribbean	*Predicted flooding using a digital terrain model (DTM), and overlaid inundation area with the location of coastal resorts.* • Created a geo-referenced database of more than 900 major coastal resort properties in 19 Caribbean countries to assess their potential risk from a scenario of 1 meter of future sea-level rise.	Inundation, Tourism, Sea-Level Rise
Impacts of Climate Change and Sea-Level Rise: A Case Study of Mombasa, Kenya[61] *(19)* University of Southampton, School of Civil Engineering and the Environment, Southampton, UK.	Kenya	*Used GIS analysis to assess the number of people and associated economic assets exposed to inundation by comparing elevation data with gross domestic production and population data to identify vulnerable assets. Used four sea-level rise scenarios and two population growth projections to determine how larger population will affect economic projections.* • Captured baseline scenario of projected economic damage from sea-level rise and storm surge in Mombasa, Kenya. • Calculated vulnerable assets based on number of people that would be exposed multiplied by the per capita GDP of the country. • Did not investigate adaptive infrastructure or other mitigation options.	Inundation, Sea-Level Rise, GIS, Developing Countries

[60] Scott, Daniel, Murray Charles Simpson, and Ryan Sim (2012). The vulnerability of Caribbean coastal tourism to scenarios of climate change related sea level rise, *Journal of Sustainable Tourism*, 20:6, 883–898.
[61] Kirshen, Paul, K. Knee, and M. Ruth. (2008). Climate change and coastal flooding in Metro Boston: impacts and adaptation strategies. *Climatic Change* 90:453–473.